OPEN LIBRARY ⊙ SCIENCE

CELL FORM AND FUNCTION

OPEN LIBRARY ◉ SCIENCE

General editor Frank Barnaby

CELL FORM AND FUNCTION

John Stares

Illustrations by Sandra Barnaby

GEOFFREY CHAPMAN

LONDON   1972

Geoffrey Chapman Publishers
35 Red Lion Square, London WC1R 4SG

Geoffrey Chapman (Ireland) Publishers
5–7 Main Street, Blackrock, County Dublin

ISBN 0 225 65853 4

First published 1972

This book is set in 10 on 12pt, Times Roman and printed in
Great Britain by A. Wheaton & Company, Exeter.

Contents

Preface

Before we can begin to understand the workings of any machine, we must study the parts of which it is built. An engineer cannot hope to understand his machines simply by standing aside and looking at them; he must take them apart, examine the components and find out how they work, both individually and combined together in a complete, functional engine. He must further study how each component is dependent on every other one.

The same is true of the science of biology, the study of living machines. The biologist cannot learn very much of the functioning of living organisms simply by looking at them. He must look inside them, at all the internal structures, and study their individual organizations and their inter-relations in a whole organism. But when the biologist looks into living things he finds that they are all made up of cells, the fundamental units of life, and so he must begin again his study of living components. He must look into the cells themselves, find out how they are made, and how their parts work as individuals and as parts of the whole cell.

In this book we shall look at the present-day knowledge of the structures and functions of living cells and their components. We shall begin with the discoveries leading to our recognition of the ubiquity of cells and then look inside a cell at the many minute internal organelles; the outer boundary which segregates the delicate internal structures from the harsh external environment; the nucleus which controls the everyday working of the cell; the machinery with which the cell manufactures the chemicals essential for life; the organelles which mount guard against invasion by harmful agents; the powerhouses of the cell which trap energy from the sun, convert it into chemical energy in the form of foods and then make this energy available to the machinery of the cell. We shall examine the way in which living cells reproduce themselves, ensuring that life on earth is a continuous process. Finally we shall look at the bacteria and the viruses, minute organisms which have structures and lives of their own. Although vastly different from the animals and plants with which we are more familiar, they may be able to tell us a great deal of the very nature of life itself. And to understand this is the ultimate ambition of the biologist.

Unfortunately there is not room to acknowledge all the people who have helped to make this book possible, but I should like to record my thanks to some of them: to Harry Thomas, who first taught me the wonders of biology; to my parents and many friends who provided constant inspiration and encouragement; to Rita Lloyd, who translated my almost illegible handwriting into a type-script; and to Nick, Robin, Nigel, Oliver, Bill and Mary, without whose help and guidance I might never have had the opportunity to write this book.

John Stares
London, 1971

1

What are Cells?

Living organisms exist in a fascinating variety of forms. But without exception, all of the members of this vast and varied living world, from the smallest microscopic ditch-water animal to a human being, from a humble yeast to a giant redwood tree, have a common theme—a fundamental structural and functional unit called a cell. A single cell may constitute a whole living organism, as in a ditch-water animal or a yeast, or may be only one of many millions which make up a human body or a tree. Some cells may have special powers, such as contraction in muscle, or the ability to trap energy from the sun in green plants, but in every case their organization is basically the same. Cells are the smallest units of living matter capable of assimilation, growth and reproduction, and only by studying cells, the way in which they are built, the way in which they work, and the complex inter-relation between their structure and their function, can we begin to comprehend the very nature of life itself.

The dawn of cell theory

Recognition that all living organisms are composed of individual units dates back only a little less than 150 years. It was in fact Robert Hooke who coined the name in 1665. Examining thin slices of cork with a simple magnifying glass, he noted that it was made up of many holes, which he compared to the rooms in which monks lived, and hence he called them cells, although what he actually saw were the dead and empty cell walls of the cork tissue, not living cells. For a long time only plants were thought to be composed of cells, but when the compound microscope was developed, professional scientists began to make detailed studies of the cellular structure of both plants and animals, and in 1839 two German scientists, zoologist Theodor Schwann and botanist Matthias Schleiden, developed a general cell theory, This stated that all living organisms, plant and animal, simple and extremely complex, are composed of cells, and that each cell can act independently but can also function as a part of a complex organism.

For some years afterwards scientists were concerned with how cells were formed. Some believed that they were formed within old cells, or that they merely crystallized from a fluid, but by looking at developing embryos it was seen that as cells grow they duplicate themselves, each cell giving rise to two daughter cells by a process of cell division called mitosis. And in 1855 Rudolf Virchow, a German scientist, confirmed the unique role of cells as vessels of living matter when he wrote *omnis cellula e cellula*, every cell is derived from a cell. Thereafter philosophical speculation on the problem of 'life' and other uncertain scientific studies and vague concepts gave way to serious experimental investigation of cells.

In the century which followed, investigators of the cell approached the subject from two fundamentally different directions. On the one hand were the cell biologists, or cytologists as they would be called now, who were concerned with developing knowledge of the microscopic anatomy of the intact cell. They began with a picture of a cell as a jelly-like blob, surrounded by a limiting barrier and containing a nucleus. As microscopic techniques advanced they showed that the cell is differentiated into numerous internal organelles, all adapted to carrying on the complex processes of the cell. And when the electron microscope was developed in the 1930s they were able to go even further and began to discover and understand the working of the cell at the molecular level. At this point the cytologists have converged with scientists of the other camp, the biochemists. Biochemists study the structure and working of the cell by first ruthlessly disrupting it and then observing the chemical activity of the chemicals and parts collected. By these methods they can trace the pathways by which the cell carries out all the various biochemical reactions which underlie the processes of life, including those which are responsible for manufacturing the substance of the cell itself. Finally they aim to be able to reconstruct the workings of the intact cell, again at the molecular level.

The results of the researches of workers in both of these areas have given us a much more detailed and complex picture of the cell than that known by the early scientists. We know that the cell is itself made up of a number of different structures or organelles, such as the mitochondria and the chloroplasts, responsible for transforming energy from the sun, storing it and making it available for use by the cell; the endoplasmic reticulum and the Golgi apparatus in which the proteins of the cell are made, stored and secreted when necessary and which are also involved in the structural stability of the cell; other small bodies in the cell which digest food, store the useful products of digestion and protect the cell from invasion by harmful substances and organisms; and the nucleus which exercises control over the day-to-day workings of the cell and hence the whole organism, and which contains the hereditary information—the genetic code—in the chemical structure of the molecules which make up the chromosomes. We also know how cells grow and multiply and how they become differentiated during development from a single fertilized egg through the embryo to a fully mature adult organism.

Size of living cells

Study of the structure of biological materials is made difficult by the fact that cells are very small and are generally transparent to light. Even today the search continues for new instruments and techniques which will provide better definition of cell structure and counteract the transparency of the cell. Before looking at some of the methods presently available for the study of cells, we shall consider the size problem.

Cell size varies over wide limits and, although some plant and animal cells are visible to the naked eye—bird's eggs some several centimetres in diameter are at first single cells and some single-celled animals such as the amoeba can be seen clearly in pond water—they are usually microscopic. Throughout this book two special units of measurement will be used to describe cell size. The micron (μ), one millionth of a metre (10^{-6} metre), is a con-

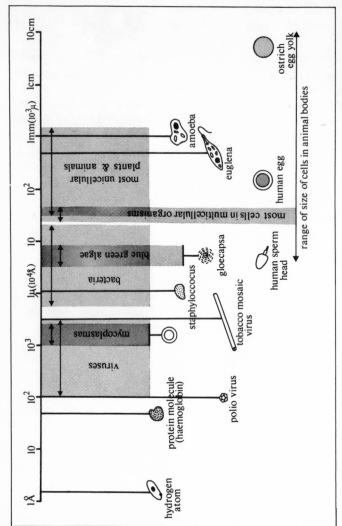

1. *Sizes of living cells compared with the size of an atom of hydrogen. The size range is so wide that a logarithmic scale has been used—if a linear scale had been used the diagram would have been one thousand million times as long. The individual cells have, of course, not been drawn to scale*

venient unit for expressing the size of cells, while the Angstrom unit (Å) which is one ten-thousandth of a micron (10^{-4} micron) is convenient when considering the sizes of some of the sub-cellular components and of the atoms and molecules which make up the cell. To give some idea of the smallness of these units, a hydrogen atom is approximately one Angstrom unit in diameter, and about one hundred billion of these would occupy an area of only one square millimetre.

The smallest free living organisms are the mycoplasmas, minute cells only about 1000 Å in diameter which are found in the soil, and are known to be the infective agents in some animal diseases. Next come the bacteria, which range in size from 5000 Å for the spherical bacterial cells, or cocci, to about 20 μ for some of the filamentous bacteria. Blue-green algae also come within this range, being on average about 10 μ in diameter. One of the largest unicellular animals, the amoeba, is about one millimetre in length, easily seen with the naked eye, although of course no internal structure is visible. Cells which make up the tissues of the multicellular organisms mostly fall into a narrow size-range of some 20 to 30 μ, and the human body is made up of some one million million of these. Human red blood cells are a little less than ten μ in diameter, and the smallest animal cells are probably about four microns across. As any object smaller than about 50 μ is invisible to the naked eye, it is clear that seeing and studying cells requires artificial techniques.

Methods of looking at cells—the optical microscope

Although the making of the first simple microscope is sometimes attributed to Galileo in 1610, no one really knows when the principle of the magnifying glass was first discovered, and it is even believed that it was known to the ancient Chinese and to ancient civilizations around the Mediterranean. What is known, however, is that all the early workers who experimented with compound microscopes, comprising more than one set of lenses to give greater magnification, were baffled by the problem of aberration—spherical aberration, or distortion of the image, and the appearance of coloured haloes round otherwise white images, known as chromatic aberration. It was not until 1845 that Charles Chevalier succeeded in making an achromatic lens which eliminated the coloured fringes, and not until a few years later that the first microscope incorporating such lenses was made in America. Modern compound microscopes are composed of a number of lenses each with several elements, which minimize both spherical and chromatic aberration.

Examination of cell structures with the optical microscope is based on differences in the refractive index of the various components. Because cells are almost transparent and the refractive indices of the components are similar, very little structure is ordinarily visible and it is thus necessary to find ways in which the contrast between the structures can be increased. One way—perhaps the most well known— is to use dyes or stains which colour some parts of the cell but not others. Some stains—called vital stains—are available which will stain living cells without killing them or disrupting their activity. Example of these are stains called neutral red, methylene blue and Janus green, the latter being of special interest because it will stain the mitochondria—bodies in the cell where energy is produced.

Vital stains have the advantage that they permit direct examination of actual living cells, but most of the early work on cell biology was done on cells which had been killed and fixed before examination, in such a way that physiological structures and chemical composition were preserved as much as possible. Towards the end of the last century it was feared that during the process of fixation the cell might be altered in some way, and that therefore the cell was not being seen as it really existed in the living state. But with modern methods this risk is very small, and it is recognized today that examination of fixed cells can yield important information on cellular structure.

Fixatives, then, are chemicals which preserve internal cellular structures in their original shape so that subsequent staining will reveal them as they appeared in the living cell. They can be either acids such as acetic acid, organic solvents such as alcohol or acetone, or water-soluble chemicals such as formaldehyde. After fixation the cells can be washed and remain stable in water. The type of stain used depends on the part of the cell to be examined. Janus green has already been mentioned as a stain for the mitochondria. The presence of deoxyribonucleic acid (DNA), the hereditary material in the nucleus which stores the information required for making proteins, can be demonstrated by using Feulgen stain, and by using a mixture of methyl green and pyronin stains it is possible to distinguish between DNA, which stains green, and the related substance ribonucleic acid, which stains red. After staining, the cells are washed in alcohol to remove excess stain, mounted on a glass slide under a coverslip in a small drop of a special gum called Canada Balsam, and are then ready for examination under the microscope.

The study of cells in whole tissues has of necessity been carried out almost entirely on fixed and stained specimens, as it is not generally possible to see cells at any depth below the surface of a tissue or organ. After fixation the tissue is embedded in hot wax, or some other substance which will solidify as it cools, and then thin sections are cut with a sharp knife called a microtome. The sections can then be stained and mounted for examination in the normal way.

Another way to increase the contrast when examining cells is to use one of the various microscopic 'tricks' which are available. These include dark-field microscopy, in which the lighting is so arranged that the specimen is illuminated from the side and very small particles are seen to reflect the light against a dark background, phase-contrast microscopy, in which very small differences in refractive index in the material can be amplified optically and made into images which can be seen, and polarization microscopy, where certain structures in the cell will show up brighter than others when viewed in polarized light. Yet another way is to use other forms of illumination. Visible light has a wavelength of between 6500 Å for red light and 4500 Å for deep violet light, and wavelengths shorter than 4000 Å are called ultraviolet radiation. Some cellular components such as DNA, RNA and proteins will absorb this radiation, so that by using a source of ultraviolet radiation instead of white light to illuminate the specimen in the microscope, it is possible to see parts of the cell containing these compounds. Mirrors and lenses made of quartz instead of glass must be used in this type of microscopy, and it is important to use suitable filters to protect the eyes against the damage which this radiation can cause. Ultraviolet radiation can also be used for another method called fluorescence microscopy. Some chemicals, when illuminated with ultraviolet radiation, absorb the radiation and then emit visible light, a process called fluorescence. Thus objects which contain such chemicals in the cell can be seen as fluorescent areas when viewed with ultraviolet radiation, and this method is useful for studying how proteins and other molecules enter or are adsorbed on to cells.

All these microscopic tricks have the advantage that they can be used to examine living cells, although methods using ultraviolet radiation have limitations in that the cells are easily damaged or killed by the radiation, and so they can only be used for short times on living tissues.

The electron microscope

We saw earlier that there was a limit to the size of objects which can be seen with the naked eye, and the same is true for objects seen with the optical microscope. This limit of resolution, as it is called, can be thought of as the smallest distance which must separate two points if they are to be seen as two. If they are closer together than the limit of resolution they will seem to merge into each other—that is, they cannot be resolved. The limit of resolution is dependent on the wavelength of the light being used to illuminate the object: the shorter the wavelength, the smaller the limit of resolution. Using white light for illumination, the limit of resolution is theoretically 2500 Å, and using ultraviolet radiation it is only decreased, again theoretically, to about 1000 Å, or 0.1 μ. Obviously in an animal cell itself measuring only 20 μ or less, very small internal structures cannot be seen using light microscopy. However, in 1933 the frontier of resolution was pushed right back by the building of the first electron microscope, and today's most powerful instrument can magnify structures up to one million times, and can resolve two points only 10 Å apart.

The electron microscope uses a beam of electrons instead of a beam of visible light to illuminate the specimen. It was known as long ago as 1924 that beams of electrons can behave as a wave in much the same way as a wave of light, and can be deflected and focused by electric and magnetic fields, just as light is deflected and focused by a lens. The wavelength of electrons moving at high velocity in a high vacuum is very small indeed—of the order of 100,000 times shorter than that of visible light—hence the very small limit of resolution of a microscope employing an electron beam as 'illumination'. Using such a microscope all parts of a cell observed with the optical microscope can be seen in much greater detail, and in addition, many other structures can be seen which were invisible with the optical microscope.

Briefly, the electron microscope works by passing a beam of electrons through a specimen. The beam is focused by magnetic fields produced by electric coils and the final image is projected on to a fluorescent screen or photographic plate sensitive to the electrons. Dark areas of the image represent areas of the cell which are opaque to electrons and light areas those transparent to electrons.

As with light microscopy, there are a number of precautions which must be taken and a number of tricks which can be applied to give a better image. Specimens must be very thin, as the penetrating power of the electron beam is very low. Sections thicker than 5000 Å are virtually opaque to electrons, and for high resolution work they must be cut no more than 100 Å thick. The usual method is to cut these sections with a glass knife produced by fracturing a glass plate. Fixation of the specimen is extremely exacting in view of the high detail attainable, and the best fixative has been found to be osmic acid or osmium tetroxide, which reacts almost instantaneously with the cellular components. This very sudden fixation maintains the structure of the cell extremely well so that it can be examined in almost the same state as when it was living. It is essential that the specimen and the path of the electron beam be maintained in a high vacuum to enable the electrons to travel with high velocity without hitting oxygen and nitrogen atoms in the air, which would deflect them and give a blurred image. The specimen must also be completely dry to avoid disturbing the electron beam.

As most biological materials are rather transparent to electrons, just as they are to visible light, stains can be used to increase contrast in much the same way that they are used in light microscopy. The stains are usually heavy metal compounds such as chromium, platinum or gold, which are dense to electrons and which are taken up to varying extents by different parts of the cell. Another technique is called 'shadow casting' where the specimen is placed in a vacuum and a heavy metal evaporated on to it from an angle, so that it is deposited on one side of the surface of any elevated structures, and forms a 'shadow' on the other. In this way the photomicrographs produced have a three-dimensional appearance, and the height of structures can be estimated by measuring the length of the shadow.

Since the first electron microscope was built there have been many developments and improvements, giving greater magnifications and better definition. One of the more recent developments has been the development of a new type of microscope called the scanning electron microscope. A fine pencil of electrons scans the surface of the specimen rather like the beam in a television tube and, as the beam scans, a series of secondary electrons is emitted from the surface. These are collected by a positively charged grid, and this signal from the grid is transferred to a television tube whose beam is scanning the screen in synchronization with the electron beam in the microscope. In this way an image is formed on the screen. Because the scanning electron microscope produces an image in much the same way as the image produced by the eye, which collects light reflected from solid objects rather like the grid collects reflected electrons in the microscope, the final photomicrograph has a remarkable three-dimensional appearance, showing the shapes of cells and their structures in fine detail.

Structure of the whole cell

What has microscopic examination revealed of the general structure of cells? The size of cells ranges, as mentioned before, from the smallest mycoplasmas which are about 0.1 μ in diameter right up to animal cells which can be as much as one million times larger. Cell size is determined at the lower limit by the smallest volume into which the minimum number of cellular components can fit and at the upper limit by the cellular activity which can be controlled by a nucleus. Another factor involved is diffusion of materials in and out of the cell and communication with the surrounding environment. If a cell's volume is above a critical limit, its surface area will not be sufficient to permit enough food materials to pass in and all the waste products to pass out. In animal cells, at least, it is found that cell volume is reasonably constant for a particular cell type, and is independent of the size of the organ of which it is a part. Animal liver and kidney cells, for example, are about the same size in a horse, a mouse and a man, and the difference in the size of the whole organ is due mainly to the different numbers of cells rather than their size.

Microscopic examination of different cells reveals a wide diversity of shape. Bacterial cells for instance may be spherical, rod-shaped or corkscrew-like or even in the shape of long thin threads. Cell shape depends on the function of the cell, surface tension, the viscosity of its internal substance, the mechanical action exerted by adjoining cells and the rigidity of its outer membrane or cell wall, and while some cells have a fixed specific shape, others can change their shape depending on the external conditions. Single-celled animals such as the amoeba can change shape by pushing out arms, or pseudopodia, which can either be used to trap food materials or can cause the animal to move along. White blood cells of animals, the leukocytes, are spherical while they are in the circulating blood, but when outside the blood system, they can also emit pseudopodia and become irregular in shape.

In the tissues, cells of both plants and animals are polyhedral because of the pressures exerted by neighbouring cells. The situation is rather like a clump of soap bubbles which can be seen to have flat surfaces where they are in contact with each other. Under these conditions the ideal shape of cells is a fourteen-sided structure which mathematicians call a tetrakaidecahedron, and if one looks at both soap bubbles and cells closely packed in a tissue it is seen that the average bubble or cell has in fact fourteen surfaces. It is important to remember, however, that cells are three-dimensional objects, and that they should be examined from as many aspects as possible to get a true picture of their shape.

Although the organization and activity of all cells are basically the same throughout the living world, microscopic examination does reveal a number of structural differences between plant, animal and bacterial cells. Detailed structure of the components of cells will be described in later chapters, but it will be useful here to consider in general terms how these types of cell differ.

Plant and animal cells appear under the microscope as transparent or translucent masses of cytoplasm, containing a large dark body called the nucleus. Just before and during cell division, a number of short rod-like bodies can be seen in the nucleus. These are the chromosomes, the genetic material of the cell, and at other times when the cell is not dividing they are diffused throughout the nucleus and are virtually invisible, even with the electron microscope. Scattered in the cytoplasm are a number of refractile bodies or cell organelles which perform specialized tasks within the cell. These include the mitochondria, the 'powerhouses' of the cell which provide its energy; the green chloroplasts—found only in plant cells—which make sugar using energy from the sun; a complex system of membranes—many of which can only be seen with the electron microscope—which act as structural components, as barriers between the cell and the external environment, and as sites for the synthesis of proteins; the lysosomes, or 'suicide bags',

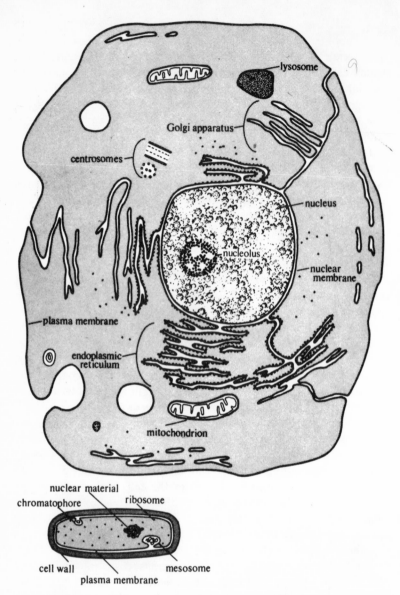

2. *Generalized eukaryotic animal cell showing the major internal organelles.* Notice that they are concentrated towards the centre of the cell, and that there is a region at the periphery of the cell which has no organelles. The smaller diagram shows that a typical prokaryotic bacterial cell has a much simpler structure. The nuclear material is not surrounded by a membrane and there are few internal organelles

which contain enzymes powerful enough to destroy the cell itself and which are involved in the digestion of food and protection of the cell from invasion by harmful substances; and various other small granules containing stores of food materials and substances made by the cell. Most of these organelles are found in the inner region of the cytoplasm, sometimes called the endoplasm. The peripheral region of the cytoplasm, called the ectoplasm, can change its consistency from that of a jelly to that of a fluid, a change which is evident in the amoeba during the extension of the pseudopodia. The most striking differences between plant and animal cells are the plant cells' rigid cell walls and large central vacuoles, which contain solutions of salts, sugars and other materials. The vacuole is not seen in young cells, but in a mature cell it may occupy as much as 90% of the total cell volume.

The feature which distinguishes both plant and animal cells from bacterial cells is the possession of a definite nucleus. Plant and animal cells, which have a nucleus bounded by a nuclear membrane are called eukaryotic cells, while the cells of bacteria and a few of the algae are called prokaryotic cells. The latter have no nuclear membrane, and their nuclear material—chromosomes—is mixed or in direct contact with the rest of the cytoplasm. There are other differences, such as the lack of mitochondria, the machinery for energy production being instead associated with the cell's boundary membrane, and these will be mentioned in later chapters. The differences between prokaryotic and eukaryotic cells are very important and fundamental as they show that from an evolutionary point of view organisms such as amoeba and other single-celled animals and plants with definite nuclei are much closer to the higher plants and animals than are the more primitive bacteria.

The biochemists' views of the cell

Understanding of the structure and working of the cell at an even smaller level of organization, that of atoms and molecules, has been the work of the biochemists using test-tubes rather than microscopes as their tools. The biochemist is interested in the chemical compounds which make up the cellular components, and the ways in which food materials are used and waste products produced.

An important technique used by the biochemists is that of cell culture, growing colonies of cells, most commonly bacterial or mammalian cells, for investigation. Growing colonies of bacterial cells is quite a simple procedure: small samples of bacteria are inoculated into a nutrient medium containing the substances necessary for cell growth, such as a source of food and energy, some mineral salts and special chemicals called growth factors, all maintained at optimum acidity and temperature. This medium can either be in the form of a liquid broth, or can be incorporated into a jelly called agar and spread on to shallow dishes called petri dishes. After incubating for a day or so, colonies of the bacteria have grown and can be collected for examination. Growing mammalian tissue is rather more difficult, and a much more complex nutrient medium is required. The most successful medium used to be a mixture of serum and an extract from chick embryos; although there are now synthetic media available, it is still generally necessary to provide some serum as a supplement for cell growth. Such techniques can of course be used by cell biologists as well, to grow cells for microscopic examination.

Once the biochemist has obtained his cells, either direct from tissues or from his tissue cultures, he proceeds to grind them up so as to separate all the cellular components. One way to do this is to mix the cells with sand and to grind the mixture together, in a homogenizer. When the cells have been disrupted, the parts are separated by spinning the suspension in a high-speed centrifuge, when the heavy parts of the cell are deposited by centrifugal force on the bottom of the tube and the lightest parts remain at the top. The first high speed centrifuge was invented in 1925 by a Swede, Theodor Svedberg, and since then instruments have been so improved that today's models can generate forces of thousands of times the force of gravity and can attain speeds greater than one million revolutions per minute. When the disrupted cell has been centrifuged, the nucleus is found at the bottom of the test tube, the rest of the cell organelles separated in bands above, and finally the rest of the cytoplasm and all the soluble substances of the cell at the top. The biochemist can now investigate the chemistry of all these fractions individually.

Chemistry of the cell

Chemical analysis of cellular components shows that most of the cell is made up of a limited number of elements; carbon, oxygen, hydrogen, nitrogen, phosphorus and sulphur. These are the inorganic building-blocks of the cell, and are joined together in various ways to form the vast range of organic substance. There are four major types of organic molecules: carbohydrates; fats or lipids; proteins and nucleic acids. Carbohydrates are known to us in everyday life as starch, sugar and cellulose, and lipids as butter and cooking oils, and both types of compound are made up only of the elements carbon, oxygen and hydrogen. They are made by plant cells using energy from the sun. Although animal cells can make carbohydrates and lipids of their own, they can only do this by starting from other carbohydrates and lipids, obtained from plants or other animals, which they break down and resynthesize. Both carbohydrates and lipids are used by cells as energy stores, and are also involved in the structure of membranes, and the cellulose cell wall of plant cells.

Proteins are made up of the elements carbon, oxygen, hydrogen and nitrogen and, in some cases, sulphur. They are built up of smaller units called amino acids joined together in characteristic sequences so that, although there are only about twenty different amino acids, they can be joined together in different orders in long chains to form thousands of different kinds of proteins.

Two classes of protein are to be found in cells, structural and soluble. Hair, wool, muscle and the gelatin in bones all contain structural proteins. They have long, fibrous molecules which link together to give stability to the tissues which they form, and which can stretch and contract, giving muscle cells their characteristic elasticity. Soluble proteins—protein chains which are coiled and folded in a complex way so that their molecules are in globular form—are called enzymes, and their job is to act as catalysts. speeding up all the chemical reactions which take place in the cell. Each biochemical reaction involves its own special enzyme, or set of enzymes, working in interconnected pathways, such that chemicals are changed slightly in each step, the product of one reaction being changed by the next enzyme in the series and so on. All these pathways of chemical reactions are known collectively as the cell's metabolism.

The fourth major group of compounds of the cell is the nucleic acids. These were virtually ignored by scientists until the 1940s, when it was discovered that deoxyribonucleic acid (DNA) stores the information needed by the cell to make proteins, and that it is also the genetic or hereditary material of the cell, being passed on unchanged from generation to generation. DNA molecules consist of long chains built up of basic units called nucleotides, of which there are four different kinds, and it is their sequence along the chain which spells out the genetic code, the message which tells the cell how to arrange amino acids to make protein molecules. Another nucleic acid, ribonucleic acid (RNA), assists in this process by carrying the message from the DNA in the nucleus to the machinery for making the proteins in the cytoplasm. When the cell divides, the DNA molecules are replicated exactly, so that each new cell has exactly the same coded information. But when cells are changed into sperms and eggs, in a process called meiosis, there is a swapping of parts of the DNA molecules, so that the new individual formed from fertilization of an egg with a sperm will not contain exactly the same coded information as its nucleus, and thus will not be exactly the same as its parents.

Scientist studying the cell still have a very long way to go before a complete picture can be built up. The general pattern of organization and function of the cell can just be discerned, but the details have still to be worked out before we can say how the living cell works.

2

The Outer Boundary

The environment inside living cells is remarkably different from that of the outside world. The concentration of simple inorganic chemicals in animal cells, for example, is quite dissimilar from that in the atmosphere or even from that in the animal's own circulating blood. This internal milieu is maintained by the cell's outer boundary, a living structure called the cell membrane, or plasma membrane.

The cell's outer boundary

Although the plasma membrane is so thin that it cannot be resolved by the ordinary optical microscope, its existence has been known for a very long time. As early as 1850, scientists recognized that the cell must be enclosed within some limiting boundary, protecting it from the external environment and, in animal cells at least, preserving its shape. It was also realized at this time that the plasma membrane was essentially a permeability barrier, permitting the entry and exit of some substances such as food materials, gases and waste products, but excluding others, and that it controls the ionic (chemical) and the osmotic (water) balance between the cytoplasm and the exterior. Later it was shown that the passage of electrically-charged particles—ions—in and out of the cell causes an electrical potential difference to be set up across the plasma membrane, and that this phenomenon has great physiological importance. It was not until the 1940s, after the electron microscope had become one of the cell biologist's working tools, that the plasma membrane was seen and photographed and its thickness measured.

Outer coats of cells

Although the plasma membranes cannot be seen, the membranes of some cells are covered by protective layers which are visible with the optical microscope. The most well known of these are the cell walls of plant cells, the structures which Robert Hooke saw in his slices of cork.

The plant cell wall is made up of polysaccharides, complex carbohydrate molecules consisting of long chains of simple sugar molecules joined together. Cellulose, the commonest polysaccharide in cell walls, is made of long chains of glucose molecules; it is arranged in a highly organized and interwoven system of minute fibres embedded in a matrix of hemicellulose, a polysaccharide molecule based on other sugars called pentoses. Other polysaccharides are also present. Lignin occurs especially in woody plants and is responsible for their great tensile strength. And in some yeast and fungal cells, the cell wall is made up of chitin, a completely different polysaccharide altogether. Chitin is a polymer of a sugar called glucosamine, and is the same substance found in the hard black shiny coats of insects.

The composition and structure of plant cell walls is related to the age and function of the cell. In young cells, in what is called the primary stage of growth, it is elastic and about one to three microns thick. As the cell matures, the cell wall increases greatly in area until the secondary stage of growth is reached and the wall is much thicker—five to ten microns thick—and more rigid, giving great strength to the cell.

Some animal cells can also be seen with the microscope to be covered with a cell coat, which is made largely of mucoprotein, a complex of protein and polysaccharide molecules. Bacterial cells also have a cell wall, and in addition a more diffuse outer layer called the capsule is sometimes present. It has been suggested by numerous scientists that in fact all cells have some form of external polysaccharide coat, but that it is only a significant structural feature in certain cells. Its general function may be to present a selective barrier on the outside of the cell, controlling which substances come into contact with the plasma membrane.

The structure of the plasma membrane

Even before the plasma membrane could be isolated and its structure investigated directly, there were many theories about its organization, mainly based on indirect evidence. In 1895 Overton was examining the permeability properties of various unfertilized egg cells, and he noted that substances which dissolve in fats and lipids—lipid-soluble chemicals—could penetrate the plasma membrane quite easily, whereas other substances which were not lipid-soluble could only penetrate the membrane and get into the cell very slowly. Overton therefore suggested that the plasma membrane was a layer of lipid, and that the lipid-soluble chemicals entered the cell by being dissolved through the membrane.

Later experiments supported Overton's theory. One of these was measurement of the electrical impedence of the cell: between 1910 and 1925 a number of scientists showed that while the contents of cells had comparatively high electrical conductivity, which would be expected if the cytoplasm contained a high proportion of substances such as proteins and inorganic mineral salts which conduct electricity, intact cells had only very low conductivity. Fats and lipids do not conduct electricity well and are said to have a high electrical impedence, or apparent resistance, and so the results of these experiments were seen as further evidence for the plasma membrane being composed of lipid. Yet another line of approach which provided support for this theory was work involving wetting the surface of cells with oil. When a small drop of oil was placed in contact with the surface of an amoeba (a species called *Amoeba dobia* was used in one such experiment), it was seen to spread out and adhere to the cell, as would be expected if the membrane contained lipids in its structure.

The next major advance in this research came in 1925, when two scientists called Gorter and Grendel provided some evidence for the actual structure of the plasma membrane. For their experiments they used membranes from red blood cells. These are easily obtained by placing red cells in distilled water, when they swell and burst, leaving the empty membranes, or red blood cell 'ghosts', behind. Using organic solvents such as acetone, they extracted all the lipid from the membranes, and then spread this lipid on the surface of water. Lipids spread on water tend to form monomolecular layers, or monolayers one molecule thick, and so it is possible to measure the area of the lipid extracted from the membranes. Gorter and Grendel found that their monolayer had a surface area twice that of the total surface area of the red blood cells they had used, so they proposed that the plasma membrane was in fact a double layer of lipid.

But the complete structure of the plasma membrane is much more than just a simple bimolecular layer of lipid, as studies by later workers showed. Experiments on the surface tension of cells gave results much lower than would be expected for simple oil-water interfaces, and it was suggested that this lowering of surface tension was due to the presence of proteins in the membrane. Observation of cells under the microscope indicated that the plasma membrane was elastic; the amoeba can stretch its membrane when it puts out pseudopodia to move along or to catch food, and red blood cells immersed in distilled water swell and enlarge to a considerable degree before they finally burst. This elasticity was ascribed to the presence of fibrous proteins—long, straight protein molecules which can be stretched and contracted, and which are similar to the fibrous proteins in muscle cells.

With all this evidence for the structure of the plasma membrane available, it was time to put forward a model for the complete structure. So in 1938 J. F. Danielli, who was working at King's College in London, suggested that the membrane was a double lipid layer covered in each side with a layer of protein. Lipids are long molecules, one end of which is polar or water-soluble while the other is non-polar or lipid-soluble, and will not dissolve in water. In Danielli's model the lipid molecules lie perpendicular to the plane of the membrane, with their non-polar ends aligned towards the centre and their polar ends pointing outwards and associated with the protein molecules. Danielli estimated that the inner lipid layer should be about 35 Å thick, while the outer protein layers should each be about 20 Å, making the total thickness of the plasma membrane about 75 Å.

Although this structure, which is now referred to as the 'unit membrane' is still recognized as being basically correct and as being the general pattern on which all membranes in the cell are modelled, more recent studies have slightly modified and extended it, for in its simplest form, it cannot account for all the properties of the plasma membrane. For example, it was discovered that lipid molecules on a water surface can form themselves into spheres, or micelles, in which the polar ends can be directed towards the centre or towards the outside of the sphere, depending on the water content of the system; there is evidence for the occurrence of this structure in the membranes of some cells.

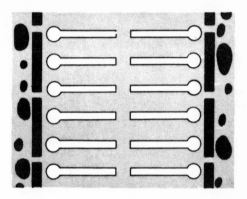

3. *Structure of the unit membrane which forms the basis of all membranes in living cells.* The lightly-shaded central core of the membrane is composed of fatty material, while the darker areas on either side are proteins. Some of the compounds in the membrane, represented by the white club-like structures, are able to associate with the fatty core at one end, and with the protein layers on the outside at the other end, and so these compounds hold the membrane together

When the electron microscope was used to examine the structure of the plasma membrane there was at last direct evidence for the correctness of Danielli's model. Photomicrographs show the membrane as two dense outer bands corresponding to the protein layers, and a central clear inner zone of lipid. And studies of many different types of cell have confirmed that the unit membrane structure is the basis, not only of the plasma membrane of all cells, but also of the membranes of the cell organelles as well. The Golgi complex, first observed in 1883 by the Italian cell biologist Camillo Golgi as a reticular structure in the cytoplasm of specially stained nerve cells, has been revealed as a complex of interconnecting membranes, all based on the unit membrane. And the mitochondria, first observed in 1910 by Gadukov, using a special microscope in which these very small particles were illuminated against a dark background, have since been shown to be composed of intricately-folded membranes, covered on the outside with another membrane. Other structures in the cell—the endoplasmic reticulum, the lysosomes and the nucleus—are all now known to be bounded by membranes, all with the basic unit membrane structure. In some cases even finer detail has been observed in membranes. For example, dark lines or bridges have been seen across the clear lipid zone, and it is thought that these may be pores in the membrane through which some molecules enter the cell or the organelle.

Although the basic structure of all membranes is the same, there are considerable variations in thickness. Generally, the thickness of the plasma membrane is greater than that of the cell organelle membranes. A Japanese scientist, Yamanoto, has even gone as far as to classify two groups of membranes on the basis of their thickness. His first group comprises the plasma membrane itself, together with membranes of the vesicles of the Golgi complex. Membranes of his second group, which are on average some 10% to 15% thinner than those of the first group, include those of the mitochondria, the Golgi complex lamella, the endoplasmic reticulum and the nuclear envelope surrounding the nucleus.

There can also be variations in the thickness of the membrane at different points in the same cell. For example, the sides and base of cells in the intestine have a thickness of about 85 Å, while at the apex of these cells, where absorption of digested food materials takes place, the membrane has a thickness of about 105 Å. All of these variations in the symmetry and thickness of membranes may be due to the presence of lipids and proteins with different functional properties: they may, however, only be due to different methods of fixation used in preparing the specimen for electron microscopic examination. More detailed study of membranes is required before this question can be finally settled.

Chemical nature of the membranes—lipids

From the early 1950s, there have been many attempts to determine the exact chemical composition of membranes. This has proved very difficult, however, as the composition can change according to the health of the organism, its diet at the time, and to other environmental factors. There are differences in the membranes of different species, and different methods of analysis can also give different results, even on samples of membrane from the same cells. Furthermore, it is extremely difficult to obtain membranes which are pure and intact, and free from any other cellular material.

Perhaps the least difficulty is found in isolating membranes from mammalian red blood cells. The cell membrane is ruptured by immersing the cells in a solution which causes them to swell and burst, and when the contents have escaped, the red cell ghosts, which still retain their characteristic biconcave shape, can be collected. But even here the final product depends on the method of rupturing the membrane and the acidity of the solution used. With a slightly acid solution, the membranes can retain up to 50% haemoglobin—the red material in the cell—while in a slightly alkaline solution, only 1% is retained. Bacterial cells can be made to burst in much the same way after their cell walls and capsules have been digested away by enzyme treatment.

Chemical analysis of plasma membranes has now firmly established that they contain protein, and lipid, together with small amounts of carbohydrate. In mammalian red blood cells, for example, the lipid fraction constitutes about 20% to 40% of the total dry weight of the membrane (depending on the method of analysis), and the protein fraction about 60% to 70%—the bulk of the membrane. Carbohydrate comprises about 5% of the total weight, and seems to be distributed between the lipid and the protein fractions. But the amount of carbohydrate appears to be somewhat lower in animal cells which contain a nucleus; in liver cells, for example, it constitutes only about 1% of the weight of the membrane.

The lipid fraction of membranes is made up of two kinds of molecules—phospholipids and steroids. A phospholipid is made up of two long parallel chains of carbon atoms joined together at one end, rather like a two-pronged fork. At the joined end, forming the 'handle' of the fork, is a phosphate group, and joined on to this

POLAR END OF MOLECULE   NON-POLAR REGION

4. Chemical analysis of cell membranes has revealed two major lipid compounds. The phospholipids (*top*) have a polar end to their molecules, which is thus able to associate with the outer protein layers of the membrane, and a non-polar region which is able to associate with the fatty core.
Cholesterol (*bottom*) interacts in some way with the phospholipids, and the combined complex adds greater stability to the whole membrane than would be possible with the phospholipid alone

a molecule containing nitrogen. In lecithin, the most common phospholipid in membranes, this nitrogenous group is called choline, but there are other phospholipids found in membranes which have different nitrogenous groups. The phosphate group and the choline together form the polar end of the lipid molecule—the end which is associated with the proteins on the outside of the membrane—while the long chains of carbon atoms form the non-polar region seen as the clear inner zone of the membrane in electron micrographs. Phospholipids form some 55% of the lipid fraction of membranes, and of this 15%—22% is lecithin. The usual proportion of lipid molecules is about 70 to 90 for every protein molecule, but the protein molecules are much larger, which explains why protein forms the bulk of the whole membranes.

The other major lipid component is a steroid called cholesterol, a non-polar compound formed of carbon atoms arranged in rings. The proportion of cholesterol varies with the type of membrane; chemical analysis has shown that there is more cholesterol in the plasma membrane than in the membranes of cell organelles such as the mitochondria. Studies with simple non-living lipid films on the surface of water have shown that when cholesterol and phospholipids are packed together in a film, they seem to interact in some way, giving the compact structure increased stability. In the plasma membrane this arrangement is probably responsible for providing the required strength and stability.

Recently more sophisticated methods of analysis have been applied to the investigation of the chemical composition of membranes. Paper and gas chromatography—methods in which mixtures of chemicals can be separated into their components and the amount of each component measured accurately, even in very small quantities—have revealed the presence of other lipids in membranes. These include small amounts of triglycerides, or fats ('three-pronged forks' of carbon atom chains with no polar 'handle') and glycolipids which are molecules containing sugars and other carbohydrates bound to lipid molecules.

Chemical nature of the membranes—proteins

The composition of the protein fraction of membranes is not nearly so well known. Danielli and his colleagues proposed in their original membrane model that the surface protein layer is in the form of sheets of expanded protein chains, rather like the fibrous proteins, and that other proteins may exist in the membrane in globular, folded form, like the soluble proteins mentioned in chapter 1. Other scientists have suggested that the inner layer of the plasma membrane, inside the cell, consists of protein, while the outer layer is probably mucoprotein—protein associated with polysaccharides. Others again have suggested that the proteins may be bound to lipid molecules in the form of lipoproteins. Probably all of these suggestions are true to some degree, but there has been relatively little success in accurately characterizing the protein fraction.

What has been established, however, and investigated quite fully, is the presence in membranes of enzymes—the biochemical catalysts involved in the complex organic reactions of the cell. In fact, some 30 different enzymes have been detected from isolated plasma membranes, in particular an enzyme involved in the passage of materials in and out of the cell. In bacterial cells, a group of enzymes involved in the final breakdown of food materials and the storing of energy is found associated with the plasma membrane, while in eukaryotic cells of animals and plants these enzymes are localized in the membranes of the mitochondria.

The ability of a protein to function as an enzyme depends critically on the molecule's three-dimensional structure—the way it is folded and coiled and the relative positions of the amino acid units which make up the protein. The enzymes in the membrane must therefore retain their three-dimensional structure, and are hence among the globular or unextended proteins in the membrane.

Functions of the plasma membrane

Membranes have two functions in living cells, although of course these functions are inter-related. The first is a structural function: the organization of the various cell organelles is based on the unit membrane structure and, as we shall see in chapter 3, these organelles provide a partition of the cytoplasm. Membranes also provide a barrier between the cytoplasm and the inside of the organelles, and of course a barrier, also, between the cytoplasm and the extracellular environment, which is fresh or sea water in unicellular organisms, and the blood, lymph and interstitial fluid in multicellular animals and plants.

The other function of membranes is their role in metabolism. All biochemical reactions of the cell are dependent on the membranes, either through the enzymes which are associated with them or through the entry and exit of substrates and waste products which are controlled by the membrane acting as a permeability controller, or organ of transport.

Many factors determine how effectively the plasma membrane performs its transport function. These include the size of the particle or molecule to be transported, the number and size of pores in the membrane structure, the living activity or metabolism of the membrane and the electrical potential difference and relative mechanical pressures across the membrane. But there are only two fundamental types of mechanism for passing materials across—passive transport and active transport.

Passive transport mechanisms

Passive transport mechanisms, also called physical processes, depend only on differences in concentration or electrical charge existing on either side of the membrane, and the membrane itself does no active work. It has already been mentioned that lipid soluble substances can easily pass across the membrane by dissolving in the lipid region. But in 1933 Collander noticed that small molecules pass through the membrane much more easily than would be expected on the basis of lipid solubility alone. He suggested that the plasma membrane was acting as a molecular 'sieve', with pores of molecular size in the otherwise continuous lipid layer. These pores are not necessarily permanent gaps, but are probably short-lived routes of entry. One idea is that the lipid layer transiently changes, or partly changes, from a layer into the micellar form, thus leaving holes. Probably the protein on the outside of the membrane unfolds in some way to line, and thus stabilize, these pores.

The important passive transport mechanisms are simple diffusion and osmosis, a special variation of diffusion. Diffusion across the membrane occurs if the membrane is permeable to solute (the dissolved substance as opposed to the solvent in which they are dissolved), the solute moving from the side of higher to the side of lower concentration. If the membrane is permeable only to solvent molecules, these will pass across the membrane from the side of lower to the side of higher concentration by osmosis. In other words, if there is an inequality of concentration, or a concentration gradient, substances will pass across the membrane in a 'downhill' direction down the concentration gradient in an attempt to achieve a state of equal concentration on either side. Across the plasma membrane in a living cell both concentration gradients and osmotic forces are considerable, as the cell contains many substances in concentrations very different from those in the environment.

Depending on the environment in which the cell normally lives, plasma membranes of various kinds of cells show marked differences in permeability. An amoeba and a mammalian red blood cell, for example, have a 100-fold difference in permeability to water. As the amoeba spends all of its life in fresh water, it is vitally important to prevent water entering the cell, otherwise it would swell up, burst and die. The red blood cell, however, is constantly bathed in plasma in which the concentration of water is similar to that inside the cell, and hence there is less need to keep water out. Placed in distilled water, red blood cells will rapidly take in water, swell up and burst.

Another force involved in passive transport mechanisms is electric potential. Different concentrations of electrically-charged ions on either side of a membrane will set up an electric potential gradient, which will act together with the concentration gradient in driving passive transport. Once again the movement is downhill; positively charged ions, or cations, will try to move across a membrane to the side where there is a lack of cations or a surplus of negatively charged ions, or anions, in an attempt to equalize the electrical charge on either side. Membrane potential can actually be measured in some large cells by inserting fine wires, called microelectrodes, with tips of one micron or less, through the membrane into the cytoplasm, or even into the nucleus; in the giant nerve cells of the squid, for example, the inside of the cell is normally 80 millevolts negative compared with the outside of the cell, or external surface of the membrane. This potential difference is called the steady or resting potential and, depending on the type of cell, it has been found to vary from between −20 and −1000 millevolts.

Membrane potential, in its role as providing one of the forces of passive transport, is of great physiological importance, particularly in animal tissues such as nerve and muscle. It enables muscle cells to contract or nerve cells to conduct nerve impulses (which are only electric currents) whenever a suitable stimulus is applied. What happens is that the region of the plasma membrane which is stimulated is momentarily depolarized or even undergoes charge reversal (the inside becomes positive with respect to the outside) due to temporary free passage of ions across it. This movement of ions causes a current to flow along the membrane to the next unstimulated region, which then itself becomes depolarized and in turn stimulates the next region and so on. Depolarization is only transient, and in just a few milleseconds the original balance is restored so that the membrane is again ready to receive another stimulus and transmit the next impulse. In other cells such as the red blood cells or plant cells, the function of the membrane potential is not so evident, and in these cases it may act only as a force in passive transport.

Active transport across membranes

Passive transport, then, always moves materials across a membrane down a concentration or electric potential gradient. But there are some cases where the movement of substances across a membrane goes uphill against the forces of passive transport, and the cell accumulates compounds which would normally diffuse away. A good example is the mammalian red blood cell which contains twenty times more potassium ions than sodium ions, while the plasma which surrounds it contains twenty times more sodium ions than potassium ions. As the plasma membrane of the red cell has passive permeability to both ions, sodium should tend to leak into—and potassium out of—the cell if passive transport forces alone were in control. But the cell is able to maintain the imbalance by constantly exuding sodium and accumulating potassium.

Many other substances are accumulated in cells against concentration gradients. Amino acid concentration within a cell can exceed that outside by a factor of 10 or more and the most spectacular example of all is that of certain marine algae, which can maintain an internal iodine concentration one million times greater than that in the surrounding sea water.

In these examples the plasma membrane is acting, not as a passive permeability barrier, but as a dynamic organ. The forces which move substances uphill against the passive forces require oxygen and use up energy, and are thus called active transport mechanisms or metabolic processes. In general they are used to transport substances which the cell requires for its biochemical reactions, but which would be unable to pass into the cell by lipid solubility or passive transport mechanisms. Size is not a limiting factor in active transport as amino acids, for example, move across the membrane more easily than do some other small molecules.

Active transport requires a number of special provisions: a source of energy; some form of carrier molecule which will pick up a molecule on one side of the membrane, carry it across and release it on the other side; and special sites on the surface of the membrane where the molecule or ion can be picked up by the carrier.

The energy source is the universal cell 'fuel' called adenosine triphosphate (ATP). Energy stored in this molecule is released when one of the phosphate groups is split off, leaving adenosine diphosphate (ADP), and this can be re-energized by the addition of another phosphate group to reform ATP. The amount of energy the cell uses for active transport is considerable; one estimate is that up to eighteen per cent of the total ATP required by the cells of the brain is used in metabolic transport processes. ATP is broken down to ADP and a phosphate group through release of energy by an enzyme called adenosine triphosphatase (ATPase)—one of the enzymes found in the protein fraction of the plasma membrane.

For active transport to work, ATP must be available inside the cell itself. This has been shown experimentally by poisoning giant nerve cells of the squid so that they could not produce ATP on their own, and then treating them with ATP. It was found that when ATP was injected into the poisoned cells, active transport activity was immediately restored to the membrane, and continued until all the ATP was used up. But application of ATP to the outside of the cell did not restore the activity. Other scientists used red cell ghosts for their experiments and found that the membrane could only transport materials in the presence of internal sodium ions and external potassium ions and again required ATP inside the cell. It was also found that ATPase activity was itself greatly increased by both sodium and potassium ions.

The plasma membrane has a number of special sites which can

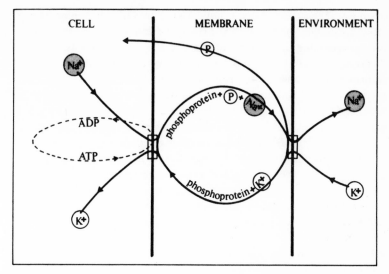

5. *Scheme for the active transport of sodium and potassium across cell mem-branes.* At a special site on the inner surface of the membrane, a carrier sub-stance, phosphoprotein, is activated by the addition of a phosphate group and energy released from the breakdown of ATP. It is thus able to pick up a sodium ion, carry it across the membrane and release it to the outside of the cell. As it releases the sodium ion, the carrier picks up a potassium ion, and carries it to the inner surface of the membrane, releasing it inside the cell. In this way the cell is able to accumulate potassium and to keep its sodium concentration low, the reverse of the situation which would occur if only passive transport mech-anisms were involved

distinguish between sodium and potassium ions, or between naturally occurring amino acids and synthetic ones with very similar but not identical structure. The number of these sites, which can only be activated from within the cell, is relatively small; on a typical red blood cell only about 1000 such sites are necessary for transport of sodium and potassium ions. It was Danielli who suggested that at these sites the ion combines with a carrier substance in the mem-brane.

The nature of the carrier has been the subject of much speculation. Various compounds have been suggested, such as phospholipids, which make up the bulk of the lipid fraction of membranes anyway, or a molecule related to the substance being carried, such as a sodium compound for carrying sodium ions. But the carrier is now thought to be a phosphoprotein—a protein with phosphate groups bound to some of the amino acids in the chain. One of the models put forward for the active transport of sodium and potassium ions suggests that the phosphoprotein is activated inside the cell by the addition of a phosphate from the breakdown of ATP to ADP, and picks up a sodium ion at the same time. It then travels to the other side of the membrane, releases the sodium ion and the phosphate group and picks up a potassium ion, which it takes across to the first side, ready to begin the cycle over again. How the carrier transports the ion across the membrane is not certain. Danielli suggested a number of possibilities: the carrier may diffuse through the membrane; it may rotate about an axis in the centre of the membrane, or it may hand its ion along a chain of similar carriers. But the true mechanism has still to be discovered.

Phagocytosis and pinocytosis

Two methods by which bulk solids and liquids can be transported into and out of cells are phagocytosis and pinocytosis, which literally mean eating and drinking. Phagocytosis is perhaps most well known in the amoeba, which uses it to trap food and take it into its cytoplasm, but it is also seen in leukocytes, or white blood cells, where it is used to trap and kill invading micro-organisms such as bacteria, and in other cells.

The mechanism is essentially the same in both phagocytosis and pinocytosis. The plasma membrane forms a small invagination enclosing a small portion of the immediate environment—a particle of solid or a small sample of fluid. Gradually the membrane closes round the sample as it is drawn towards the centre of the cell, until the invagination closes up completely, forming a vacuole enclosing the sample. Finally this vacuole is released into the cell, to float free in the cytoplasm. Technically the contents of the vacuole are still outside the cell, in much the same way as food materials in an animal's stomach are still outside the animal's body until they are absorbed through the stomach wall. But the contents can be digested by secretions released through the vacuole membrane, after which the digested products can be absorbed into the cytoplasm through the membrane and the waste products can be passed out of the cell by reverse phagocytosis, the vacuole travelling back to the

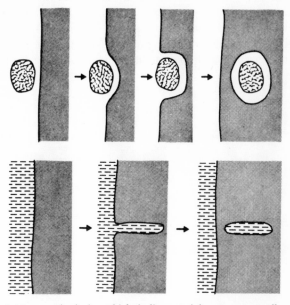

6. Two methods by which bulk materials can enter cells are phagocytosis (*top*) and pinocytosis (*bottom*). Phagocytosis is responsible for the entry of solids and pinocytosis for the entry of liquids, but the mechanism is the same in both cases. An area of the plasma membrane forms a small invagination which gradually deepens and engulfs either some solid or some liquid. Eventually the invagination closes over the solid or the liquid so that a vacuole is formed within the cell, and the plasma membrane has closed behind it

plasma membrane, fusing with it and then opening to release its contents.

Phagocytosis in reverse is in fact used by specialized cells, such as cells of the pancreas, for secreting substances made by the cell. Pancreas cells produce enzymes, and inactive forms of these are packaged in small membrane-bound vacuoles, which then travel to the plasma membrane and release their contents to the rest of the body. The enzymes must be secreted in an inactive form, or they would start digesting the substance of the cell itself before they could be secreted.

Specialization of the plasma membrane itself may be related to function in some cells. Cells of the intestine have numerous finger-like villi, or projections, at their surface, and this arrangement increases the surface area of the cell by a factor of 20 or more. This enlarged surface area probably provides attachment for enzymes related to digestion and also facilitates absorption of the products of digestion.

3

The Nucleus

When cells, living or dead, are examined with the microscope, perhaps the most evident visible structure is a large dark dense body within the cytoplasm—the nucleus. This structure used to be thought of as the most important part of the cell, but in the light of what we now know about the functions of all the other cellular organelles and their interdependence, this view is open to challenge.

But although perhaps not strictly the most important structure within the living cell, the nucleus is certainly its central control room, governing and directing all the processes by which energy is made available to the cell, organic compounds are manufactured by the cell's other machinery in the cytoplasm, and the general health of the cell is maintained in good order. For the nucleus contains the cell's genetic material, the detailed instructions for the production of nucleic acids, proteins and the other regulatory substances necessary for the cell life, growth and proliferation, all coded in the chemical structure of the deoxyribonucleic (DNA) molecules which form the chromosomes, and from which messages can be passed out to the cytoplasm and translated. It is one of the functions of the nucleus to keep this genetic material intact and to pass it on unchanged from generation to generation. Since it was first observed in 1835 by a Scottish botanist in plant cells, the nucleus has been the subject of more intense study than any other cellular component.

Life-cycle of the nucleus

The life-cycle of any cell can be thought of as having two distinct phases: a resting stage or interphase, and a phase when the cell is in the process of division. The length of this cycle varies enormously in different cells. In some single-celled organisms cell division is very frequent, and under optimum conditions they will divide once every 20 minutes, while in others, such as mammalian nerve cells, the interphase may be almost as long as the life of the whole organism of which the cell is a part. But however long the cycle, the appearance, structure and role of the nucleus during each of the two phases is very different. During interphase, the nucleus is concerned with directing synthesis—protein synthesis particularly in the early interphase just after the cell has divided, and the synthesis of DNA in late interphase just before the cell undergoes another division. During this phase also, the synthesis of some of the other cellular organelles such as the ribosomes takes place. But during the division phase, the nucleus performs very different functions, being mainly concerned with the transfer of the genetic material to the new daughter cells. In fact the nucleus does not even exist as the same entity at this time, as the membranes surrounding it disintegrate, and profound changes in its internal structure take place. All of these changes will be discussed in detail in chapter 6, while this chapter will be concerned with the structure and role of the nucleus during interphase.

External appearance

The existence of the nucleus was first suspected as far back as 1700, when van Leeuwenhoek, a Dutch draper and amateur scientist, using simple home-made lenses, detected refractile bodies in the centres of the red corpuscles of salmon blood. Although he did not realize it at the time, these structures must have been nuclei (the red blood cells of salmon, in common with those of some other animals but unlike those of man, do contain nuclei). But it was the Scottish botanist Robert Brown's observations in 1835 that nuclei can be seen quite generally in the cells from several tissues of flowering plants which really began the immense interest in this very special part of the cell.

The shape of the nucleus is sometimes related to that of the whole cell, although this relationship is by no means universal. Typical shapes are spheroid, discoid, or even irregular; for instance, in some white blood cells, or leukocytes, it appears to be multilobate— a series of bodies, all joined by thin threads and arranged in the shape of a horseshoe. But in the early stages of cell division of most cells, it is usually spherical. The size of the nucleus is also variable, but in general it appears to be related to the size of the cell as a whole.

The number of nuclei in a cell can also vary in different organisms and tissues. Most cells are mononucleate, but binucleate cells— some liver and cartilage cells, for example—are known, and in some cases there may be more than two per cell. Some cells of the bone-marrow have been observed with up to 100 nuclei, and in a specialized type of tissue called syncytium, where a number of cells seem to have fused together and lost their individual plasma membranes so that the whole mass of cytoplasm is surrounded by only a single membrane, the nuclei may be even more numerous. Striated muscle in animals is made up of this type of tissue, and some muscle fibres contain several hundred nuclei.

The position occupied by the nucleus in a cell is generally characteristic for a particular cell type. In embryonic or young cells it is almost always in a geometric centre of the cell, but it commonly becomes displaced as the cell develops and differentiates or reserves of materials are formed in the cytoplasm. In plant cells with large central vacuoles, it is seen pressed up against the cell wall with the rest of the cytoplasm.

Internal organization

In eukaryotic plant and animal cells the nucleus is surrounded by a double membrane, too thin to be resolved by the optical microscope, called the nuclear membrane or nuclear envelope. In prokaryotic cells such as the bacteria, this structure is naturally absent, and this is the main distinguishing feature of the two classes of cell. Each of the two membrane layers of the nuclear envelope is of the same general lipid and protein structure as the plasma membrane, and is about 90 Å thick. The space between the two layers is normally some 140 Å wide. The inner membrane appears to be a continuous sheet, and is intimately associated with the internal matrix of nucleoplasm, or nuclear sap, while the outer layer seems to be a continuation of the endoplasmic reticulum, and is perforated at a number of points with pores large enough to permit the passage of fairly large organic molecules, thus providing a connection between the nuclear sap and the cytoplasm. The actual number and distribution of these pores differs in different types of cells. Unlike the plasma membrane, which can withstand considerable strains and stresses, the membranes of the nuclear envelope can survive only very minor damage.

In the interphase nucleus, the nuclear sap appears to be a colourless, translucent to transparent, homogeneous fluid or weak gel with much the same consistency and characteristics as the cytoplasm. In 1869, Miescher, working in Switzerland, was able to isolate nuclear material from pus on hospital bandages. He called this material nuclein. Later on he worked with salmon sperm cells, because he had noted from previous observations that these cells were composed almost completely of nuclei, and he was able to show that the nuclein he extracted contained the elements phosphorus, carbon, oxygen, hydrogen and nitrogen. Later still, when nuclein was found to be acidic, the name of this material was changed to nucleic acid. Today, with the aid of the electron microscope, it is possible to see that the nuclear sap is composed of granules, which chemical analysis has shown to be essentially nucleic acids, proteins and some inorganic salts of magnesium and calcium.

But although apart from these granules very little structure is evident in the interphase nucleus, even with the electron microscope, there had been evidence dating from the middle of the last century that the nuclear sap was not homogeneous. Scientists studying cells in the division phase of their life-cycle had observed that when the cell is about to divide, small bodies become visible in the nucleus, and it was further discovered some years later that these bodies could be stained with special basic dyes to stand out against the surrounding matrix. These bodies were thus called chromosomes (Greek: *chromos*, colour; *soma*, body). In the interphase nucleus it is extremely difficult to see these chromosomes, even with the high magnification available in the electron microscope; even with special staining techniques, only a network of chromosomal material, called chromatin, is visible in the form of fine filaments. The reason for this is that in the interphase state the DNA of the chromosomes is in a highly hydrated and thus swollen form, and has a refractive index very similar to that of the surrounding nuclear sap.

On the basis of their staining reaction, two types of chromatin can be distinguished: euchromatin and heterochromatin. The name of the latter was coined in 1928 by E. Heitz to describe the chromatin which stained more deeply than the euchromatin during interphase and less deeply during cell division. This different staining reaction is in fact due to quantitative differences in the chemical composition of the two types; while the euchromatin contains a standard amount of nucleic acid at all times, the heterochromatin contains more nucleic acid during interphase and less during division. Improved methods of isolation and purification of chromosomes from dividing cells and chromatin from cells at interphase has enabled detailed analysis of this material to be carried out and have shown that it consists of four major types of compound: small proteins, with low molecular weights, called histones; varying quantities of other types of proteins; and the two nucleic acids RNA and DNA.

Molecular structure of the nucleic acids

The discovery of the structure of the nucleic acids is a fascinating story of scientific investigation, beginning with the initial isolation of nuclein by Miescher. In 1880 a German chemist, Emil Fischer, found that the nucleic acids contained two heterocyclic bases (carbon and nitrogen atoms arranged in rings) called purines and pyrimidines. Some years later Kossel was able to describe these bases in more detail and demonstrated the presence of two purines, called adenine and guanine, and two pyrimidines, cytosine and thymine. Then, in 1910, a Russian-born biochemist, Levene, added his contribution by isolating a five-carbon sugar molecule called ribose from the material. Later he found that another type of nucleic acid contained another five-carbon sugar called deoxyribose, and he went even further by suggesting that the phosphorus which Meischer had found was present in nucleic acids as phosphate groups. These act as links between complexes of sugar molecules and bases, so that the whole molecule is a chain of alternate sugar molecules and phosphate groups, with a base joined to each sugar. Much later, in the 1950s, Todd in Britain succeeded in making artificial structures of this kind in the laboratory, and he found that they were identical to the natural material obtained from nucleic acids. Names were given to these building units: a purine or pyrimidine linked to a sugar molecule was called a nucleoside, and nucleosides attached to phosphate groups—the real building blocks of the nucleic acids—were called nucleotides. DNA molecules, therefore, are really long chains of nucleotides. In the course of this investigation, both Kossel and Todd received the Nobel Prize for their efforts.

By the 1940s it was commonly recognized that it was the DNA molecules in the nucleus which formed the genetic material of the cell and accounted for a large proportion of the chromosomes, but the exact three-dimensional structure of DNA was not known, nor was the way in which it could carry the immense amount of genetic information needed by the cell. In 1947, however, E. Chargaff made an important discovery which was to lead to answers to both of these problems. From very precise chemical analysis of DNA he found that purines and pyrimidines were present in equal proportions, and, more important still, that the amount of adenine was exactly equal to that of thymine, and the amount of cytosine to that of guanine. Three years later M. Wilkins, working in London, obtained valuable results from X-ray studies of purified fibres of DNA from animal cells. He found that the purines or pyrimidines were arranged at regular intervals along the long chain of the molecule, 3.4 Å apart, and that there was another structural regularity which occured every 34 Å. He suggested that the DNA molecule was not a straight structure, but that it was twisted into a helix, with one full turn of the helix occurring every 10 nucleotide units, every 34 Å. He also performed some density measurements on his molecules, and the results from these indicated to him that each molecule of DNA consisted of more than one nucleotide chain.

But the exact special structure was still not worked out, and in the next few years many eminent scientists put forward their ideas on how many nucleotide chains were involved, and how they were twisted together. Then in 1953 James Watson and Francis Crick, from their laboratory in Cambridge, told the world of their now famous double helix structure for the DNA molecule. And together with Wilkins, these two scientists were awarded the Nobel Prize for their brilliant work.

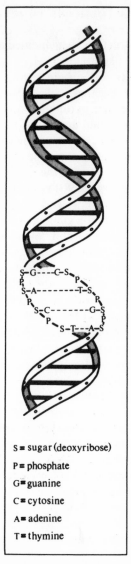

S = sugar (deoxyribose)

P = phosphate

G = guanine

C = cytosine

A = adenine

T = thymine

7. *Double helix structure of the DNA molecule* showing how the two long cha
of nucleotides are joined and twisted together. Each chain is composed of
alternate sugar molecules and phosphate groups, with the bases attached to the
sugars. A base on one chain can only join with a base on the other chain according to the rule that adenine can only join with thymine and guanine only with
cytosine

The double helix structure, which fitted all the evidence then available, looks rather like a ladder which has been twisted into a helix. The two uprights of the ladder are formed of deoxyribose sugar molecules joined together with phosphate groups to form a 'backbone'. The purine and pyrimidine molecules attached to the sugars are turned inwards and are joined to purine and pyrimidine molecules on the other chain, forming the rungs of the ladder. But the joining of the purines and pyrimidines at the middle of the rungs is very specific. For all the parts of the molecule to fit together in an orderly structure which fits the X-ray data from Wilkins, a purine must be paired to a pyrimidine, and furthermore only a thymine can pair with an adenine, and a guanine with a cytosine. Thus there are equal amounts of adenine and thymine, and of guanine and cytosine, which explains why Chargaff's results were so important. If one nucleotide chain, or side of the ladder, has a sequence cytosine-thymine-cytosine-adenine-guanine, then the other chain on the other side of the ladder must have a sequence guanine-adenine-guanine-thymine-cytosine, and the specific pairing referred to above occurs all the way along the molecule. Another implication of this structure is that one chain is upside down with respect to the other, as though with two halves of a ladder, one half must be inverted before it will fit to the other.

The structure of the other nucleic acid molecule RNA is different in three basic respects from the DNA structure. The sugar molecule is of course ribose instead of deoxyribose, and instead of thymine as one of the two pyrimidine bases there is another one called uracil. The third difference is that the RNA molecule consists of only one chain of nucleotides or one half of a ladder, so that there is very little helical structure and the molecule is usually coiled or folded in some way.

Nucleic acids as the genetic material

It has been suspected since the beginning of the century that the nucleic acids of the chromosomes carried the genetic information of the cell. In 1903 Walter Sultan and Theodor Boveri suggested that the hereditary particles or genes which endowed the cell with the ability to make a particular protein, or the whole organism with some specific characteristic, were located on the chromosomes, and in 1913 A. H. Sturtevant put forward the idea that the genes were organized in linear sequence along the chromosomes. The first actual demonstration that DNA was involved in cell heredity came in 1944 from the work of O. T. Avery and his co-workers. They took some strains of bacteria called pneumococcus which had fully developed 'smooth' capsulated cells, and extracted the DNA from them. Then they injected this DNA into another strain of pneumococcus which had 'rough' cells with no capsule, and found that when these treated cells reproduced some of the offspring cells were 'smooth'. This result indicated that the DNA had been incorporated by the 'rough' cells and used to direct their synthetic machinery, and that the 'smooth' characteristic had thus been inherited. But just how this inheritance worked could not be understood until after Watson and Crick had produced their model and the mechanism in which the DNA itself was synthesized had been worked out.

Watson and Crick themselves were careful to point out that the very structure of the DNA molecule indicated a method for its synthesis and replication in the cell. If the two polynucleotide strands of the molecule could be made to unwind and split apart, and a supply of the separate nucleotide building blocks was available, these could attach to the uncovered purines and pyrimidines of each existing nucleotide chain according to the rules of adenine joining only to thymine and guanine only to cytosine. The phosphates of the nucleotide could then join up to make two new strands, and two DNA molecules would be formed, each consisting of an original strand and a new strand of nucleotides but each whole double helix molecule being an exact replica of the original. With some slight modifications, this hypothesis was later confirmed by experiment. RNA is synthesized in the nucleus by a process similar to DNA replication. In this case the DNA unwinds and splits as before, the nucleotides line up on just one of the uncovered DNA strands and are joined together by the phosphate groups, and then the new strand is released and the DNA reforms into its original helix. Of course, for RNA synthesis, uracil and not thymine nucleotides line up and pair with the adenines on the DNA chain. In all these processes, the DNA molecule is said to act as the template for DNA replication or RNA synthesis.

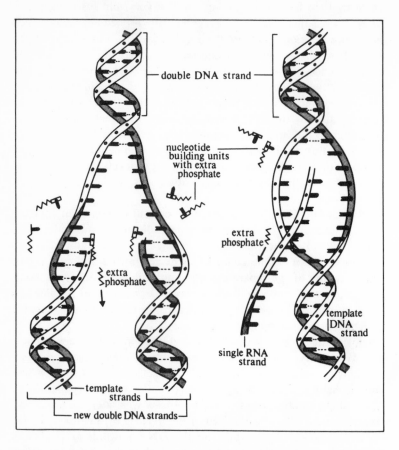

In the diagram:
- double DNA strand
- nucleotide building units with extra phosphate
- extra phosphate
- extra phosphate
- single RNA strand
- template DNA strand
- template strands
- new double DNA strands

8. Replication of DNA (*left*) involves unwinding of the double helix so that nucleotides can line up opposite the exposed bases, an adenine nucleotide lining up on an exposed thymine base, and so on. The combination of the nucleotides into a long chain is facilitated by the presence of the extra phosphate, which is split off as the chain is formed. Each new DNA molecule is composed of one chain of the original molecule and one newly formed chain which is complementary to it, so that two new identical molecules are produced.

In RNA synthesis (*right*) only one chain of the DNA molecule acts as the template for the formation of the new nucleotide chain, and the two DNA chains reform into the original molecule

The genetic code and protein synthesis

In the structure, and mechanism of replication and synthesis, of DNA and RNA is the means by which the genetic information of the cell can be stored, used to direct the protein synthetic machinery of the cell, and passed on unchanged to the next generation. What the nucleus has to do is to tell the machinery in the cytoplasm the order in which amino acid molecules must be arranged and joined together to make a specific protein, and these instructions are carried in the sequence of the different bases in the DNA of the chromosomes, the specific order of these bases in a short section of the DNA being the coded instruction for a particular amino acid. The problem of just how many bases are required to specify one amino acid worried scientists for some time, as somehow just four bases must provide enough information to specify 20 different amino acids. If the code were one base on the DNA chain specified one amino acid, only four amino acids could be specified, and if the code were two bases for one amino acid, then only 16 (4×4) could be specified— better, but still not enough. With a code of three bases for each amino acid, however, there are 64 $(4 \times 4 \times 4)$ possibilities—more than enough. Direct evidence for this code (called the triplet code) of three bases in the DNA chain telling the cytoplasm to insert one amino acid in a protein chain was provided by Crick. He took some DNA and altered it by adding or removing bases and then examined the proteins which were synthesized from it. When only one base was removed or added, protein synthesis was drastically altered, but when one removal was combined with one addition not far away along the molecule, the message was restored and synthesis proceeded as normal. When three removals or three deletions were made close together, the message was again unaltered, ample evidence that the code operates on sequences of three bases along the DNA molecule.

When the DNA molecule replicates, therefore, this coded information is also replicated because of the strict rules governing which bases can pair with each other in forming a new molecule. And synthesis of RNA, using the DNA as the template and governed by the same rules, provides the means for passing the information out of the nucleus to the protein synthetic machinery in the cytoplasm. Three types of RNA are made by the nucleus; messenger

The Genetic Code

UUU	Phe	CUU	Leu	AUU	Ile	GUU	Val
UUC	Phe	CUC	Leu	AUC	Ile	GUC	Val
UUA	Leu	CUA	Leu	AUA	Ile	GUA	Val
UUG	Leu	CUG	Leu	AUG	Met	GUG	Val
UCU	Ser	CCU	Pro	ACU	Thr	GCU	Ala
UCC	Ser	CCC	Pro	ACC	Thr	GCC	Ala
UCA	Ser	CCA	Pro	ACA	Thr	GCA	Ala
UCG	Ser	CCG	Pro	ACG	Thr	GCG	Ala
UAU	Tyr	CAU	His	AAU	Asn	GAU	Asp
UAC	Tyr	CAC	His	AAC	Asn	GAC	Asp
UAA	stop	CAA	Gln	AAA	Lys	GAA	Glu
UAG	stop	CAG	Gln	AAG	Lys	GAG	Glu
UGU	Cys	CGU	Arg	AGU	Ser	GGU	Gly
UGC	Cys	CGC	Arg	AGC	Ser	GGC	Gly
UGA	stop	CGA	Arg	AGA	Arg	GGA	Gly
UGG	Trp	CGG	Arg	AGG	Arg	GGG	Gly

KEY

U Uracil
C Cytosine
A Adenine
G Guanine

Ala alanine
Arg arginine
Asn asparagine
Asp aspartic acid
Cys cysteine
Gln glutamine
Glu glutamic acid

Gly glycine
His histidine
Ile isoleucine
Leu leucine
Lys lysine
Met methionine
Phe phenylalanine

Pro proline
Ser serine
Thr threonine
Trp tryptophan
Tyr tyrosine
Val valine

9. The information coded in the structure of the DNA of the chromosomes for the synthesis of proteins is the genetic code. Each sequence of three bases in the DNA molecule tells the protein synthetic machinery to add a particular amino acid to a growing protein chain. As there are 64 possible triplet sequences and only 20 amino acids, each amino acid is specified by more than one triplet. Three of the triplets, instead of coding for an amino acid, tell the synthetic machinery that it has come to the end of a protein chain, and these are marked as 'stops' in the code

RNA, transfer RNA and ribosomal RNA. Messenger RNA actually carries the information from the DNA to the cytoplasm. The sequence of bases in the messenger RNA is in fact complementary to that in the DNA—a sequence cytosine-adenine-guanine in the DNA will appear as guanine-uracil-cytosine in the messenger RNA—but the information content is still the same. Transfer RNA is coiled up in such a way that it has two special regions exposed. One of these regions is a series of three bases which form the code for a particular amino acid, while the other is able to attract and hold that amino acid. The function of this RNA is to bring the amino acid molecule to the messenger RNA strand and hold it there by temporarily joining to the messenger with its three bases, according to Chargaff's rules. In this way the code on the messenger RNA strand is read, piece by piece, by the transfer RNAs, and the amino acid molecules are joined together in the correct sequence to form a protein.

The nucleolus

Ribosomal RNA is the material which will eventually form the bulk of the ribosomes, the small particles in the cytoplasm where the synthesis of proteins takes place. Ribosomes are formed in the nucleus in a structure called the nucleolus which is associated with a particular site on the chromosomes called the nucleolar organizer, and it is the DNA in this region of the chromosome which forms the template for ribosomal RNA synthesis.

The nucleolus is the only definite body which is ordinarily visible in the interphase nucleus. It is a small dense body which shows up particularly well after staining and there are often two or more in each cell, rounded, oval or irregular in shape.

Nucleoli were first observed in 1781 by Fontana, and by the end of the nineteenth century they had been studied sufficiently for a relationship between their size and the protein synthetic activity of their host cells to be postulated. They are seen to be small or even absent in cells exhibiting little synthesis of proteins, such as sperm cells, while in cells such as those in the pancreas producing digestive enzymes, where protein synthesis is a prominent feature of the cells' activity, they are correspondingly large. Chemical analysis of isolated nucleoli shows that they are made up mostly of protein—about 80% is an average figure—although little is known of their enzyme content. They are also very rich in RNA, but they do not stain with Feulgen stain and hence contain no DNA.

With the optical microscope, nucleoli appear structurally homogeneous, although small corpuscles and vacuoles which form clear zones within the dense matrix are sometimes noted. With the electron microscope more detail can be seen, and the matrix is made up of long fibrils some 50 to 80 Å long and granules 150 to 200 Å in diameter. Both the fibrils and the granules are made up of RNA and protein, in a complex called ribonucleoprotein, and careful study has revealed that the fibrils coil and fold up and eventually clump together to form the granules. And it is the granules which will eventually form the ribosomes which will be passed out into the cytoplasm to carry out their work in protein synthesis. Thus cells which manufacture large amounts of protein need large numbers of ribosomes, and hence need large nucleoli to make them. There is also evidence that the nucleolus is the site for storage of other RNAs before they are passed into the cytoplasm. In bacterial cells where there is no nucleolus or nuclear envelope, the ribosomes are released into the cytoplasm immediately after they are made in the appropriate region of the DNA.

Proteins of the nucleus

The proteins of the chromatin, histones and the other larger proteins, are found in association with the DNA, forming a complex called nucleoprotein. Most studies on the nature and role of the proteins of the nucleus have been confined to the histones, and their biological function has intrigued biologists for a long time. They are found associated with DNA along the whole length of the chromosome, and it used to be thought that the histones were the important part of the chromosome with a direct part to play in heredity, the role of the DNA being ignored. But now that the true importance of the DNA has been confirmed, other functions have been ascribed to them. One idea is that histones act as a sort of 'chromosomal glue', binding together the individual genetic units of DNA. They may also have a role in stabilizing the DNA; it is known that the histone-DNA complex is more resistant to damage by heat than DNA alone, and that the protein in nucleoprotein can partially protect DNA from radiation damage.

But all the evidence we have at present indicates that the really important role of the histones lies in repressing the genetic activity of the cell. Histones are able to react with DNA in a specific way, preventing it from acting as a template for RNA synthesis and thus preventing the transfer of genetic information out into the cytoplasm. There is also some evidence that some of the other proteins in chromatin act as derepressors, neutralizing the effect of the histones, but the process is by no means clear. Possibly the proteins together act as modifiers of the activity of the nucleus, and thus do have some effect on the hereditary mechanisms of the cell.

Giant chromosomes

As we have seen, in the interphase nucleus the DNA is usually diffused throughout the cytoplasm as fine filaments, and is not visibly organized into chromosomes until the cell is about to divide. The appearance of chromosomes during cell division will be discussed in chapter 6. There are, however, some very important exceptions to this rule. The salivary glands of many insects contain specialized non-dividing cells, in which the chromatin is permanently in the form of visible 'giant chromosomes'. Such chromosomes are called 'giant' because they contain about one thousand times as much DNA as normal chromosomes, and can be up to three hundred times as long. These cells, instead of growing to a certain size and then dividing, continue to grow and become very large without going into the division phase. The chromosomal constituents are duplicated many times, with the duplicated molecules of DNA remaining attached to the original chromosome. These chromosomes are sometimes called polytene chromosomes, and the cells which contain them—the most well known are the salivary glands of the fruit fly *Drosophila*—are important because they provide a very useful and convenient means for biologists to study chromosomes direct, and to determine which parts of them are responsible for which hereditary characteristics.

Polytene chromosomes can be seen to have a characteristic pattern of lateral bands along their length, the pattern being different from fly to fly. The four chromosomes in the *Drosophila* together have about five thousand bands. When these chromosomes are stained with Feulgen stain, only the bands take up the stain, the interband regions hardly taking up any stain at all, which indicates that DNA is present in much greater quantities in the bands than in the other regions. In fact the many strands of DNA are lying parallel in the interband regions, but are coiled up in dense loops in the bands themselves. This coiling can be shown by stretching the chromosomes, using micromanipulators fitted with microdissection needles. When these are inserted into the cell, the chromosomes can be stretched to more than 10 times their normal length. In normal cells where there is only one DNA strand per chromosome, this form of coiling only takes place when the cell starts to divide.

The bands in polytene chromosomes have been extensively studied by geneticists, who have found it possible in many cases to locate, by microscopic examination, the exact position of a particular gene containing the coded information for a particular protein or

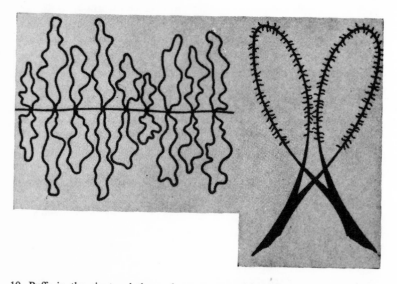

10. Puffs in the giant polythene chromosomes of fruit fly cells are composed of many loops similar to the one shown at right. The fine hairs are strands of messenger RNA being synthesized on the DNA chromosome strand, so the presence of the puffs shows that this gene is sending out instructions to the cytoplasm. In different tissues and at different times, puffs are seen on different regions of the chromosome, showing that different genes are active. The diagram at left shows a lampbrush chromosome, found in amphibia. Each loop is also seen with the electron microscope to have many fine hairs of RNA attached to it

characteristic. Although the chromosomes lie very close together and appear to be one rather than two, they are in fact paired as in normal cells, one being inherited from the mother and the other from the father. If one of these chromosomes is deficient or damaged in any way, some of the bands will be missing on just one of the paired chromosomes. Such an abnormality can easily be seen with the optical microscope, and related to any deficiency, such as the lack of a particular protein, in the fly.

Salivary gland chromosomes are of special interest for another reason. Some regions of polytene chromosomes are modified at characteristic points. They become enlarged and, in extreme cases, puff out to form large rings round the chromosome, called chromosome puffs. Staining these puffs shows that, in addition to DNA and proteins, they contain large amounts of RNA, and they are in fact regions of the chromosome where RNA is being synthesized, the degree of enlargement being related to the rate of synthesis. There is a characteristic change in the puffing pattern as the insect develops from a grub to a mature insect, showing that different genes are being activated, and that the nucleus is sending out messenger RNA instructions for different proteins to be made as the various processes of development occur.

Another type of giant chromosome is the lampbrush chromosome of developing eggs of frogs and other amphibia, so called because it resembles the type of brush used for cleaning the glass chimney of old-fashioned paraffin and kerosene lamps. These are long strands about two hundred Å long, with many symmetrical loops clearly visible on each side of the chromosome with the optical microscope. These loops have a fluffy appearance, and the 'fluff' can be shown to be RNA, so that this is another case where the synthesis of RNA can be followed and the activity of genes monitored.

4

Cytoplasm and the Organelles

The cytoplasm can be thought of as the most important region of the cell, for it is within this region that most of the metabolic and biosynthetic functions and work of the cell take place. This is the cell's true internal milieu, where food is broken down and processed, waste products are formed and the enzymes and nucleic acids necessary for energy production and synthesis are localized.

The nature of the cytoplasm

Because of the limitations of optical microscopy, early observations of the cell showed the cytoplasm as a homogeneous, amorphous region, in which were embedded the cell organelles such as the mitochondria. This homogeneous material was called the ground, or fundamental cytoplasm, or the hyaloplasm, and, although it has general properties in common throughout all organisms, the exact chemical nature of its proteins and RNA in any one species seems, at least in mammals, to be distinct.

Then towards the end of the last century it was discovered that portions of the ground cytoplasm in some cells had differential staining properties; in fact these regions stained with the same basic dyes which could be used to stain the nucleus. They were thus recognized to have a role in biosynthesis and were hence called the ergastoplasm, from the Greek *ergazomai*, meaning to elaborate or transform.

One of the most characteristic properties of the cytoplasm is its

60

ability to change from a liquid-like substance, or sol, to a semi-solid jelly-like substance, or gel. Physical chemists call materials which undergo this type of sol-gel transformation colloids, and the colloidal properties of cytoplasm have been studied by a number of techniques. For example, minute steel spheres can be implanted in the cytoplasm of cells by microsurgery, and by drawing these spheres through the cytoplasm with magnets it has been possible to measure its viscosity. Such experiments have shown that the relative viscosity of the plasma sol is quite low—in fact only two to ten times greater than that of water—while the plasma gel is more rigid, shows elastic properties, and can actually be broken by application of a critical force. Other workers have demonstrated that the plasma gel can be liquified under high pressure.

In the living cell regions of both sol and gel cytoplasm can be distinguished. In the main, the gel is located at the periphery of the cell near the plasma membrane, a region which is generally free from granules or other organelles and inclusions. On the other hand, the cytoplasmic sol in the interior of the cell contains all the cytoplasmic organelles, which will be discussed in this chapter.

The vacuolar system of the cell

The most obvious feature of the sol region of the cell is a complex system of membranes which pervade the cytoplasm and form numerous compartments and sub-compartments. The development of this system, generally called the cell vacuolar system, varies in different cell types and according to the cell's differentiation and function, but in all cells it seems to separate the cytoplasm into two parts, one contained within the system, and the other—the cytoplasmic matrix proper—outside it.

The cell vacuolar system was not discovered until techniques for electron microscopy of intact cultured cells and thin sections became available, and further advances of our knowledge of this system were only made by applying biochemical analysis and cytochemical techniques to studying the specific components individually. It is made up of the nuclear envelope—the membrane surrounding the nucleus—the endoplasmic reticulum and the Golgi complex, all membrane-bound organelles responsible for vital cellular functions.

But the vacuolar system has a number of functions as a whole, some of which are well known facts while others are still only tentative hypotheses awaiting confirmation. The first function is support. By dividing the fluid content of the cell into compartments, the membranes of the vacuolar system provide supplementary mechanical stability for the colloidal structure of the cytoplasm. The membranes also act as a barrier regulating the passage of materials between the inner and outer compartments or between the components and the cytoplasmic matrix, in much the same way as the plasma membrane regulates the passage of materials into and out of the cell.

Because of this exchange function, the membranes of the vacuolar system probably have special enzymes such as ATPase to facilitate transport. But in any case there are a number of enzymes, mainly involved in the metabolism of lipids, associated with the membranes of the endoplasmic reticulum, and the membranes themselves provide a large surface area on which the metabolic reactions may take place. Also associated with the membranes are enzymes for protein synthesis, and when the proteins are made for export from the cell—cells in the pancreas, for example, make digestive enzymes which are discharged into the gut to break down food—the vacuolar system is involved in their storage and eventual secretion. The system does not seem to be so involved when the proteins are manufactured for internal use by the cell.

Endoplasmic reticulum

In 1945 electron microscope pictures of cells revealed the presence of a reticulum or network of strands and membranes in the cytoplasm. Subsequent, more detailed, studies showed that these strands were in fact connected, vesicle-like bodies forming a complex branching network of membrane-bound cavities called cisternae in the inner region of the cytoplasm. As this region of the cytoplasm is sometimes called the endoplasm, the new structure was appropriately named the endoplasmic reticulum. It does in fact extend from the nuclear envelope all the way through the cell to the plasma membrane and in some cells it can be seen to be an actual extension of the plasma membrane connecting it to the outer membrane of the nuclear envelope (the nuclear envelope is made up of two membranes). This in turn suggests that the nuclear membrane is continuous with the plasma membrane, perhaps providing a direct communication route between the nucleus and the extracellular environment.

The form of the endoplasmic reticulum varies with the cell type. It may be made up of vesicles, small tubes or large flattened sacs or cisternae, and it is frequently folded to form layers of membranes. The membranes themselves are rather thinner than the plasma membrane, being only about 50 to 60 Å thick, but they nevertheless have the same general organization—a double layer of lipid material covered on each side with protein. The endoplasmic reticulum is often absent in egg cells and in embryonic or undifferentiated tissues, but it becomes evident and increases in complexity as the cells mature. But whatever the variation in form, two distinct types of endoplasmic reticulum can be recognized: rough, with a granular appearance; and smooth, lacking granules. The granular appearance of rough endoplasmic reticulum is due to the presence of large numbers of small RNA-rich particles called ribosomes attached to the sides of the membranes. Smooth endoplasmic reticulum, of course, has no granules associated with it, and in cells with little or no endoplasmic reticulum ribosomes are to be seen floating free in the cytoplasm.

The actual amount and type of endoplasmic reticulum in a cell varies considerably with age, with the function of the cell and with prevailing extracellular conditions. A simple, smooth reticulum is found in cells engaged in lipid metabolism such as adipose tissue or cells of the adrenal cortex. But in cells which are actively engaged in protein synthesis—pancreatic cells, for example—there is a highly-developed system, consisting of large cisternae covered with ribosomes. In liver cells, rough endoplasmic reticulum is seen in the ergastoplasm regions, while the regions rich in glycogen (a polysaccharide of long chains of glucose molecules, but joined together in different ways from those which make up cellulose) are seen to contain tubular, smooth reticulum. There is, however, continuity between both types at a number of points. It has been estimated that the total surface area of endoplasmic reticulum contained in one millilitre of liver tissue is about eleven square metres, some two-thirds of this being the rough granular type.

In plant cells, smooth endoplasmic reticulum develops near the periphery of the cell, where the cellulose cell wall is formed, which implies that it has a role in polysaccharide metabolism. And its presence in the glycogen-rich regions of liver cell is a further implication for this relationship. But biochemical studies have shown that the enzymes necessary for polysaccharide metabolism are not present either in or on the membranes and that the reticulum is not concerned with these reactions. The membranes do, however, incorporate the enzymes necessary for the metabolism of steroids—special lipids with a chemical structure related to cholesterol—and especially those steroids which function as hormones, and the reticulum as a whole has an important function in the storage of lipids, including steroids, in vacuoles in animal cells and in the plastids of plant cells.

But the chief role of the endoplasmic reticulum is in the synthesis and storage of proteins. During the 1950s Palade concluded from his studies on cells that there was a fairly close correlation between the amount of rough endoplasmic reticulum in a particular cell and the quantity of protein manufactured for export from that cell. Some of the main protein-secreting cells of the animal body are the plasma cells, which produce antibodies, the fibro blasts, which produce a fibrous protein called collagen, and the cells of the pancreas, which produce and secrete digestive enzymes used in the breakdown of food materials in the gut. In all of these cells there is a highly-developed rough endoplasmic reticulum consisting of large cisternae covered with large numbers of ribosomes. But in cells where proteins are produced primarily for internal use, such as muscle cells, there is little or no endoplasmic reticulum, and ribosomes can be seen floating free in the cytoplasm.

Ribosomes and protein synthesis

The ribosomes are the sites where the cell makes its proteins and they are among the most fundamental of the sub-cellular particles. They are found in virtually every type of cell, not only on the endoplasmic reticulum or in the cytoplasm, but also in the nucleus and nucleolus where they are made. They are between 100 and 150 Å in diameter, divided into two sub-units, one larger than the other, and viewed under high magnification they have a slightly angular profile. Chemical analysis of isolated ribosomes shows that they contain a high proportion of ribonucleic acid.

A group of ribosomes concerned with the synthesis of a specific protein is arranged in a functional group called a polyribosome, or more simply a polysome. This looks like a string of ribosomes linked to a thread of a special RNA called messenger RNA, each ribosome being arranged so that the upper sub-unit appears to be on top of the messenger RNA strand, while the lower sub-unit seems to be slung below it. Messenger RNA is made in the nucleus as a copy of the coded information contained in the DNA of the chromosomes for the arrangement of amino acid molecules which will form the protein. As the ribosome travels along the messenger RNA strand it 'reads' the coded information. Amino acid molecules are brought to the ribosome on the back of special carriers called transfer

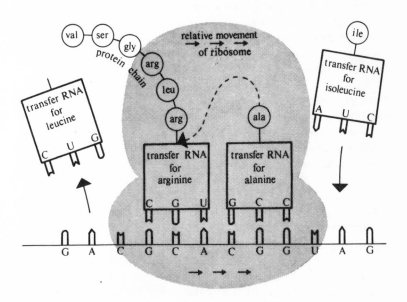

11. Synthesis of proteins in the cytoplasm takes place on the ribosomes moving along a strand of messenger RNA. Transfer RNAs line up along the messenger strand according to their triplet of bases, and give up their amino acids to the growing protein chain. They are then released, the ribosome moves along one place, and the next transfer RNA lines up. The abbreviations for the bases and the amino acids are the same as those used in the Genetic Code Table, page 54.

RNAs, each of which has a coded region of its own which can recognize and lock on to a part of the code on the messenger RNA. When the transfer RNA is so locked into position, it gives up its amino acid molecule to the growing string of the protein, after which the ribosome moves along one place and reads the next part of the messenger RNA, and another transfer RNA with its amino acid is locked into position. By the time the ribosome has reached the end of the messenger RNA thread, a complete strand of protein has been built up. It then detaches from the messenger, releases its protein and is free to join on to the beginning of another messenger RNA strand and begin the process over again. With the help of the electron microscope and special staining techniques it has been possible to photograph this process. The messenger RNA strand appears rather like the central shaft of a feather, with the partially completed protein strands radiating out on each side. At one end the strands are short, while at the other where the ribosome has almost finished its work, the protein strands are much longer.

The Golgi complex

If the proteins made on the ribosomes are to be used by the cell itself, they are stored in the cytoplasm until needed. But if they are to be secreted from the cell they penetrate the membranes of the endoplasmic reticulum, where they are segregated from the rest of the cell. When the time comes for them to be exported from the cell, another part of the vacuolar system becomes involved. This structure is called the Golgi complex.

The Golgi complex is a system of minute canals and vesicles located near the nucleus of all cells except bacteria and mycoplasmas, and is particularly prominent in secretory cells. It was discovered in 1883 by the Italian biologist Camillo Golgi while he was experimenting with techniques for staining cell preparations for microscopic examination. He fixed preparations of the nerve cells from a barn owl and a cat with formaldehyde and then stained them with silver compounds, which caused the Golgi complex to stand out black against the surrounding cytoplasm. But the complex is difficult to see in living cells as its refractive index is very similar to that of the ground cytoplasm; despite the Italian's demonstration, its existence was questioned for a long time. Eventually the electron microscope revealed that the canal-like structures consist of a system of vesicles of various sizes bounded by membranes and arranged in a somewhat parallel pattern. Some recent evidence indicates that a Golgi complex may be present in some plant cells, but in any case there are similar structures which botanists call dictyosomes.

Detail revealed by the electron microscope shows that the Golgi complex consists mainly of large, flattened sacs, or cisternae, with walls of double layers of lipoprotein membranes, clusters of vesicles some 600 Å in diameter intimately associated with the cisternae, and large clear vacuoles at the edge of the complex. But the main characteristic is the absence of ribosomes, which means that the Golgi complex has no ability to manufacture proteins of its own. A recent theory is that it is a modification and extension of the smooth endoplasmic reticulum.

The relationship between the Golgi complex and cell secretion was first postulated in 1914 by Cajal, and is now largely accepted. Recent research has shown that it plays an important role in packaging proteins made by the cell ready for export. Proteins made for use outside the cell are first stored in the cisternae of the rough endoplasmic reticulum, after which they accumulate in the Golgi complex. Here some carbohydrate molecules are added to the protein, and large numbers of the molecules are finally wrapped together within a single membrane. This package may then migrate from the Golgi complex to the edge of the cell, where the membrane fuses with the plasma membrane and the contents are released to the outside—a sequence of events which is in fact a reverse of phagocytosis.

This mechanism is employed to make the mucous part of the saliva which is secreted from the salivary glands, and of course to liberate the digestive enzymes of the digestive glands into the gut. A pancreatic cell makes two protein-digesting enzymes, called trypsin and chymotrypsin, which are released in this way. In an inactive form, called zymogen, they accumulate in the Golgi complex as granules. The portion of the Golgi containing these granules then buds off from the main complex and migrates to the plasma membrane. Once outside the cell, the zymogen can change to the active forms of the enzymes which then start their work of digesting food materials.

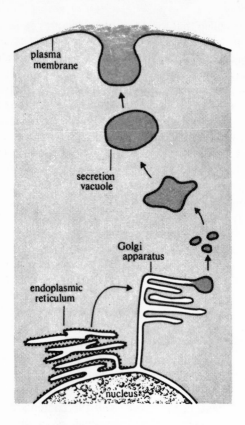

12. Proteins which are synthesized for export from the cell accumulate in the Golgi complex. Small membrane-bound sacs then bud off from the Golgi complex and gradually coalesce into larger vacuoles, which migrate to the plasma membrane, fuse with it, and release their protein contents. It is by this process—really a reverse of phagocytosis—that cells release digestive enzymes into the gut

The lysosomes

Not all of the enzymes packaged up by the Golgi complex, however, are destined for the extracellular world. Some are wrapped into membranous sacs which are then released to float free in the cytoplasm. These single-membrane-bound vacuoles digest food materials within the cell and have a role in the defence mechanisms of the cell; they are known as the lysosomes.

The discovery of the lysosomes was effected partly by the biochemists and partly by the cell biologists with their electron microscopes. In 1949, Christian de Duve, working in his laboratory in Louvain in Belgium, was studying the amounts of enzymes in various fractions of homogenized rat liver cells, after they had been separated by centrifugation. During his investigation he noticed that the quantity of one enzyme, called acid phosphatase because of its ability to act on phosphate esters, varied considerably in apparently identical experiments, and even increased after the fraction had been left standing for a few days. He also noticed that when he homogenized the cells only gently he obtained much less of the enzyme than when he subjected the cells to vigorous homogenization. De Duve suggested that during both vigorous homogenization and during ageing, this enzyme was being released from some source within the cell. This was the first clue to the existence of the lysosomes.

Furthermore, it was known that cells which had a high content of the acid phosphatase could also contain many phosphate esters, apparently unattacked by the enzyme. So it was suggested that the enzyme was segregated from the rest of the cell inside some organelle, and that only when the membrane surrounding this organelle became permeable during ageing, or broken during vigorous homogenization, was the enzyme released.

Lysosomes are very difficult to identify, even with the electron microscope, as they have no characteristic shape or internal structure, and they are of the same order of size as the mitochondria—between 0.25 and 0.80 μ in diameter. They can be distinguished from the mitochondria in that they have no internal folds of membranes, but depending on their state of activity at any one time, their size, shape and internal inclusions may vary. However, in 1955 Novikoff clearly identified lysosomes in rat liver cells with the electron microscope using a special method of electron staining. It is now thought that they occur in all animal cells and that they are formed by the endoplasmic reticulum and the Golgi complex. Vacuoles similar in appearance to the lysosomes have been seen in some plant cells, but as yet their biochemistry has not been fully investigated and they cannot be definitely identified.

Biochemically the lysosomes are characterized by their unique enzyme content. These enzymes—more than a dozen have been positively identified—function primarily as catalysts in what are called hydrolytic reactions where large organic molecules are broken down by the addition of water. They are capable of digesting proteins, nucleic acids and polysaccharides, and such highly active enzymes must of necessity be segregated in sacs from the main body of the cell, or they would start digesting and destroying the cell itself. They are called lysosomes because of their ability to destroy, or lyse, substances, and because they could destroy the cell itself, they are often referred to as 'suicide bags'. But as long as the membranes are intact, no damage can be done.

Lysosomes have many different functions in living cells. When food materials are taken into the cell in the form of large complex molecules by phagocytosis, they must be digested or broken down into simpler molecules which the cell can utilize. When a lysosome encounters a vacuole formed by phagocytosis (such vacuoles are sometimes called phagosomes), the membranes of the two fuse together to form a digestive vacuole. Within this vacuole, the large molecules are digested into smaller compounds which pass through the membrane into the cytoplasm, by one of the processes described in chapter 2. The vacuole then migrates to the edge of the cell and any undigestible material is excreted from the cell by reverse phagocytosis.

Materials broken down by the lysosome enzymes need not be food from outside the cell. Worn out parts of the cell can be digested in vacuoles, the products of the digestion sometimes being released to the cytoplasm and recirculated as food. Even whole dead or worn-out cells can be completely broken down by lysosomal action. White blood cells or leukocytes have an important role as scavengers, engulfing dead tissues or harmful invaders of organisms such as bacteria and destroying them with the help of the lysosomes, and in this way the lysosomes are acting as part of the defence mechanism

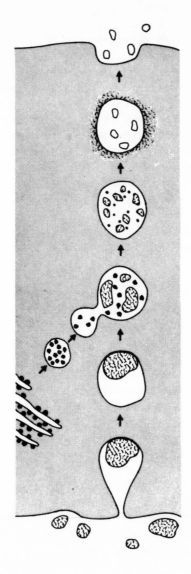

13. A major function of the lysosomes is in digestion of food materials within the cell. Vacuoles formed by phagocytosis fuse with lysosomes, when the corrosive lysosome enzymes are released into them. When the enzymes have broken down the food materials, the useful products are absorbed into the cytoplasm through the membranes of the vacuoles, and the debris is ejected from the cell by reverse phagocytosis

of the cell or the organism. But sometimes the lysosome's defence role can go tragically wrong, and can result in considerable damage. A disease called silicosis—common among miners who are constantly in the presence of, and hence inhaling, silica dust—may result from silica particles accumulating in the lysosomes in the cells of the lung. This accumulation of hard abrasive particles damages the delicate lysosomal membranes and causes the corrosive enzymes to leak out into the cytoplasm and start digesting the substance of the cell itself.

Other agents in excessive quantities can also be responsible for affecting the permeability of the lysosomal membranes. Vitamin A, for example, is known to cause the membranes to become unstable and to leak. But cortisone and hydrocortisone have the opposite effect of stabilizing the membrane, a property which may account for the well known anti-inflammatory properties of these substances used as drugs.

Lysosomes are also involved in the embryonic development of organisms; in fact they have a role right at the beginning in the fertilization of egg cells. Lysosomes in the spermatozoa release their corrosive enzymes to digest away some of the substance surrounding the egg, thus facilitating penetration by the sperm and subsequent fertilization. Another well-known example comes from the development of the frog. When the tadpole develops into the frog, its tail is 're-absorbed' as a result of the action of lysosomal enzymes in the tail cells and the molecules so released are used by the rest of the organism as food materials. Cell death of this type by the action of lysosomes is now known to be a normal part of the process by which many tissues and organs are remodelled during embryonic development.

And yet another process in which healthy living cells are broken down by the lysosomes and the products recycled—a process known generally as autodigestion or autophagy—is initiated during starvation of the organism. Some of the healthy tissue is engulfed in vacuoles and digested, and in this way the cell is able to use some of its own material to provide essential substances which are no longer available from the outside, without causing irreparable damage to the whole cell.

Plant cell vacuoles

From time to time the cell is seen to contain a number of small vacuoles or granules within the cytoplasm. These can be either the remains of vacuoles formed by phagocytosis or pinocytosis, or storage vacuoles, containing reserves of lipids or carbohydrates, and are usually fairly insignificant structures. But plant cells contain a huge central vacuole which can sometimes occupy most of the cell.

The plant cell vacuole is bounded by a single membrane called the tonoplast, which may be formed by expansion of the cisternae of the endoplasmic reticulum. In young plant cells only numerous small vacuoles are present, some of which may not be visible with the optical microscope, but as the cell grows these small vacuoles enlarge and fuse together until in a mature cell there is only one vacuole which may occupy as much as ninety per cent of the whole cell volume.

The tonoplast appears to have special permeability characteristics. Sugars, mineral salts and other substances are present in the vacuole in high concentrations, and the high osmotic pressure developed within the tonoplast may help to keep the cytoplasm pressed hard against the cell wall, thus giving rigidity to the cell. Pigments may also be present within the vacuole, giving the cell a characteristic colour; colours of many flowers are due to the presence of these pigments in the vacuoles of the petals. The vacuole may also function as a dustbin for the cell, containing substances which the cell no longer needs or cannot use.

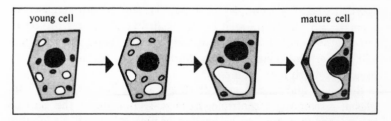

14. *Development of the central vacuole in plant cells.* Minute fluid-filled vacuoles in young cells gradually enlarge and coalesce until, in the mature plant cell, there is just one large vacuole

Centrosomes, flagella and cilia

Adjacent to the nucleus, lying just within the cytoplasm of all animal cells and some plant cells are bodies called centrosomes. They were first seen by van Beneden in 1880 in cells of certain parasites called cephalopods. He called them *corpuscles polaires*, and since then the electron microscope has been used to examine them in great detail.

Centrosomes are made up of two separate bodies called centrioles, each of which is about 0.2 μ in diameter and rod-shaped. In electron microscope pictures they look like pairs of hollow cylinders, rather like pairs of open bean-cans at right angles to each other. And in cross-section, they are clearly seen as bundles of nine small rods arranged round the outside of the cylinder, each rod consisting of three fibres.

The centrioles seem to serve as a centre of activity during mitosis, and in fact are particularly evident at this stage of the cell's cycle. As cell division starts, the centrioles divide and proteinous fibres radiate out from the centriole and form an assembly called the aster. The two pairs of centrioles then migrate to opposite sides of the nucleus, the proteinous fibres between them elongating to form the spindle, the structure on which the chromosomes will line up and be separated during mitosis. In plant cells which do not contain centrioles, the early stages of formation of the spindle are less well defined, and it does not become organized until other mitotic changes, such as the disappearance of the nuclear membrane, occur.

It is interesting that this arrangement of groups of rods is also seen in other structures in cells. Some cells are able to swim about by beating thread-like organs called flagella and cilia. The tail of a spermatozoa, for example, is a flagella, and many single-celled animals and plants are seen to have hair-like structures projecting outside their cells. Flagella are usually present singly and are about 150 μ long, while cilia are much shorter, usually five to ten microns long, and much more numerous.

Both cilia and flagella grow from a granule, just within the cytoplasm, resembling a centriole. In cross-section under the electron microscope they are seen as bundles of nine rods, each made up of two fibres, with an extra double fibre in the centre. This similarity to the centriole suggests that they could have evolved from the same

15. *Internal structure of a flagellum.* The arrangement of nine rods, each composed of a pair of
fibres, with another pair of fibres in the centre is
similar to that found in cilia and centrosomes

primitive structure. Chemical analysis of flagella and cilia show that
they are basically protein—70%–84% of the dry weight—with
smaller amounts of lipids and traces of carbohydrates and nucleic
acid material. Bacterial flagella and cilia are somewhat different,
in that there is no arrangement of rods, the threads being composed
of globular molecules about 40 Å in diameter packed into a long
fibre.

5

Mitochondria and Chloroplasts

All living organisms, plants and animals, depend ultimately for their energy on the sun. The harnessing of the sun's energy, its storage and subsequent release in the form which the cell can utilize, is the work of very important organelles, the chloroplasts and the mitochondria. In fact these organelles are often referred to as the 'powerhouses' of the cell.

Energy for living organisms

Trapping the energy from the sun is exclusively the work of green plants which carry out a process of photosynthesis. They use simple inorganic chemicals—carbon dioxide and water—and synthesize organic carbohydrates. In plants, this process takes place in the chloroplasts, organelles containing a pigment called chlorophyll which is essential for trapping light energy and which gives plants their characteristic green colour.

Releasing the energy stored in the carbohydrates is the work of the mitochondria which, unlike the chloroplasts, are found in almost all cells, the only exception being bacteria, red blood cells and a few yeasts. In the mitochondria, a compound with two carbon atoms, a 'halfway stage' in the breakdown of the food materials such as carbohydrates and lipids is finally broken down to carbon dioxide and water with the release of energy which is stored in the form of ATP, the universal 'cell fuel'. ATP is then released from the mitochondria into the cytoplasm to be used to provide energy for all the complex biochemical reactions of the cell.

The mitochondria

Mitochondria were first observed towards the end of the last century as granular structures in striated muscle cells. They were originally called bioblasts, but the name mitochondria was first coined by Benda in 1897 (Greek: *mitos*, thread; *chondros*, a grain). Then in 1900 Michaelis succeeded in staining them with Janus green. This was an important advance because it showed that mitochondria could bring about a colour change in a dye, a colour change similar to that observed when iron compounds are changed from green to yellow in a chemical reaction called reduction. For many years biologists had been studying oxidation and reduction reactions in extracts from cells without considering the possibility that the enzymes responsible were localized in cellular particles. A few years later Kingsburg actually suggested that the mitochondria were the sites of cellular oxidation, but his suggestion was largely ignored. And it was not until 1934, when the first isolation of rat liver mitochondria established the possibility of their direct biochemical investigation, and into the 1940s when improved techniques of centrifugation facilitated the isolation of cellular particles and the electron microscope and other techniques were available for studying them, that the gap between the biologists and the biochemists was finally bridged and the true role of the mitochondria was established.

The structure of the mitochondria is now reasonably well confirmed. Electron microscope pictures show them as double-walled sacs, the walls being membranes similar in structure to the plasma membrane (although they are biochemically quite different). The outer membrane is about 60 Å thick and quite smooth, while the inner one, also of about the same thickness, is full of folds and convolutions called cristae, projecting into the interior of the organelle. Within this membrane is a space, or lumen, filled with a gel-like matrix, which is homogenous except for a few fine filaments and which contains a high proportion of soluble proteins and other substances. The cristae do not project right across the lumen, so the internal matrix is continuous. The space between the inner and the outer membranes is between about 60 Å and 80 Å wide and is remarkably constant for many types of mitochondria; it presumably contains organized layers of protein.

If the mitochondria are made to swell and burst, and the membranes are then stained and examined under high magnification with the electron microscope, even more detail can be seen. Studded all over the inner surface of the cristae are structures shaped like knobs on the ends of sticks. These studs, which represent enzymes involved in the final stages of ATP formation, occur at approximately 100 Å intervals, and estimates put their total number at some 10,000 per mitochondria. Normally they are concealed within the membrane and cannot be seen in section, and it is only when the mitochondria are burst and the membranes specially stained that they become visible.

The size of mitochondria varies from about 0.5 μ to 1.0 μ in diameter, and they can be as much as seven microns long. But depending on the functional state of the cell, it is possible to see very thin mitochondria (0.2 μ diameter) or very thick ones (up to 2.0 μ). The number of mitochondria in a cell also varies enormously. Some cells possess only a single mitochondria, some simple fungi have only a few and some sperm cells have about twenty. At the other end of the scale, liver cells contain about one thousand mitochondria, which constitute some thirty per cent of the total protein content of the cell, and certain animal egg cells and amoeba can have as many as 300,000. In some diseases such as cancer, the number of mitochondria often diminishes. But with all this variation, there is an increasing accumulation of evidence to suggest that each cell type contains a specific characteristic number of mitochondria, and that this number is preserved by division after the cell has divided in mitosis. Generally mitochondria are fewer in plant cells, as some of their functions are performed by the chloroplasts.

The shape and distribution of the mitochondria within the cell also vary within wide limits, and both of these factors must be considered in relation to their role as energy suppliers. They are generally filamentous or granular in shape, but they may be cylindrical, as in kidney or liver cells, or slab-like, as in muscle cells. They can also change shape; long thin particles can swell at one end in the form of a club, or can flatten out like a tennis racket. Mitochondria may also become vesicular by the appearance of a central clear zone. The filamentous mitochondria have been seen to fragment into granules, which sometimes reunite into the filament. But all these changes are dependent upon the state of the cell at the time.

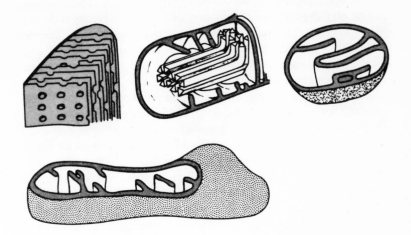

16. *Variations in internal structure of mitochondria.* The lower diagram shows the general internal structure—a continuous outer membrane, with an inner membrane highly folded into cristae which project into the organelle. Structures above show how the arrangement of the cristae varies in different cells with different functions

Variation in internal configuration and surface area of the cristae is even more profound, and is more dependent on the function of the cell. Those cells which respire slowly, such as liver cells, have mitochondria with relatively few cristase but a large amount of matrix. Muscle cells, on the other hand, respire rapidly, have little matrix in their mitochondria, but have very many cristae which almost fill the lumen. Close examination of the cristae themselves shows yet more variation. In liver cells their infoldings form an accordion-like pleated instructure. In kidney cells, which respire a little more rapidly, mitochondria have large numbers of disc-like cristae arranged like a stack of coins. In the mitochondria of the flight muscle tissue of the dragonfly, which respires very rapidly, the inner membrane is invaginated to form thousands of microvilli, which vastly increase the surface area of the cristae. In blowfly mitochondria, the lumen is almost completely filled with regularly-arranged, sheet-like cristae, each perforated at regular intervals so that the holes are exactly aligned in the structure. These perforations' may be an adaptation to permit easy diffusion of oxygen, ATP and other compounds through these rather large mitochondria. But the internal structure, although taking widely different forms in different cells, is nonetheless more or less constant in cells of similar type and those performing similar functions.

Distribution of mitochondria throughout the cytoplasm is generally fairly uniform, but there are many exceptions to this rule. The technique of time-lapse photography, where living cells are photographed under the microscope at short intervals of time, has shown that they can move around within the cell, this movement being less pronounced while the cell is in the process of mitosis than when it is in the intervening resting stage or interphase. During mitosis they are concentrated near to the spindle and on division are distributed almost equally between the two new daughter cells. In some diseased states, they are seen to accumulate round the nucleus or in the peripheral cytoplasm, a region which is normally devoid of organelles. But under normal conditions, the movement and distribution of mitochondria is again related to their role as energy suppliers, and they tend to move to places in the cell where energy is needed at that time. In some cells they can take up permanent positions near to the region where most energy is required. In the rod and cone cells of the retina in the eye, for example, all the mitochondria are to be found in the portion of the upper segment of the cell which is receiving the light. Mitochondria of the cells of the kidney tubule are intimately related to the infoldings of the plasma membrane, presumably to supply energy for all the active transport mechanisms for the compounds recovered by the kidney during the formation of urine.

The work of the mitochondria

Understanding of all the biochemical reactions and processes which take place within the mitochondria requires quite detailed knowledge of biochemistry, and their description in any detail is outside the scope of this book. The following description of how the mitochondrion makes energy available to the cell will thus of necessity be brief and simple. To give some idea of the complexity of the system, a single mitochondrion contains more than seventy different enzymes and coenzymes which work together in an orderly sequence of reactions to carry out its work, in addition to numerous other substances such as co-factors and metals which are essential for the mitochondrion's function.

Briefly, the work of the mitochondria is to make adenosine tri-phosphate (ATP) by joining together adenosine disphosphate (ADP) and phosphate in a process called phosphorylation, energy being stored at the same time in the chemical bond between ADP and the third phosphate. Water and carbon dioxide are formed as by-products of this process. This synthesis of ATP is brought about by a combination of two very important biochemical pathways, each with its own set of enzymes. Biochemists have a number of names for each of these pathways, but they will be called here the citric acid cycle (often called Krebs cycle, as it was an eminent biochemist called Krebs who suggested the pathway from all the isolated pieces of evidence available at the time), and the electron transport chain.

The three major food materials—proteins, carbohydrates and lipids—are all broken down by enzymes in the cytoplasm to acetate—a small molecule containing just two carbon atoms and some hydro-gen atoms. Acetate is then bound to a coenzyme called coenzyme A to make acetyl coenzyme A, which penetrates the membranes of the mitochondria. Once inside, the acetate is split off and enters the citric acid cycle by combining with a molecule containing four car-bon atoms to make a molecule of citrate, a six-carbon compound. A complex series of reactions involving about ten enzymes now follows, and carbon is removed as carbon dioxide at various points until a four-carbon molecule is left, and this then combines with another acetate molecule to begin the cycle over again. At several points in the cycle of reactions energy is liberated from the carbon com-pounds in the form of energy-carrying hydrogen atoms or pairs of electrons (both are in effect the same), and these are ready to be passed along the electron transport chain—another series of reac-tions, with its own set of enzymes. When they reach the end of the chain, the hydrogen atoms combine with oxygen to form the other by-product of the complete process, water. At three points along this chain of reactions energy is lost from an energy-carrying hydrogen atom or pair of electrons and this energy is used in the synthesis of ATP from ADP and phosphate. Thus the electron transfer chain is said to be coupled to phosphorylation. Because oxygen

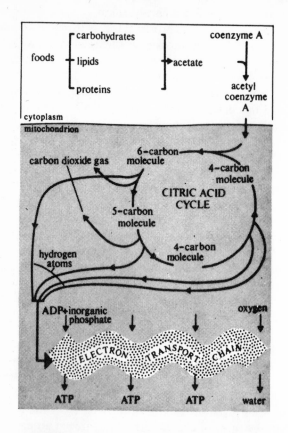

17. *Biochemical processes in mitochondria.* Foods are broken down in the cytoplasm to acetate—a molecule with two carbon atoms—and are combined with co-enzyme A. This compound then passes into the mitochondrion, where it combines with a four-carbon molecule to make a six-carbon molecule. This is then broken down by the enzymes of the citric acid cycle, releasing two molecules of carbon dioxide and energy in the form of hydrogen atoms. In the electron transport chain, the energy from the hydrogen atoms is stored in ATP molecules, and the hydrogen eventually combines with oxygen to form water

is involved in this sequence of events, this form of energy release is called aerobic respiration. But some organisms can release energy from food materials without using oxygen, in a process called anaerobic (without air) respiration. Yeasts, for example, do this by means of alcoholic fermentation, the process being carried out by enzymes in the cytoplasm, not isolated in mitochondria.

The enzymes responsible for both the citric acid cycle and the electron transport chain in mitochondria have been studied very carefully. It is now almost certain that those of the citric acid cycle are located in solution in the matrix, probably near to, or even on the folds of the cristae, but only loosely attached. The enzymes of the electron transport chain, on the other hand, are located in the membranes of the cristae themselves. In bacterial cells which contain no mitochondria, all of the enzymes of both pathways are closely associated with the plasma membrane, sometimes localized in special areas called mesosomes.

Nucleic acids in the mitochondria

Very recently our understanding of the structure and function of mitochondria took a great leap forward when it was discovered that they contain their own DNA, have their own machinery for the synthesis of proteins and can undergo division. The presence of DNA in mitochondria was hinted at between 1956 and 1957 when it was shown that mitochondria of cultured cells stained with Feulgen stain, the dye used to stain DNA, but most scientists were unwilling to attach any importance to this find, believing that all the DNA in a cell was localized in the nucleus. Even the presence of some DNA in mitochondria after cell homogenization and centrifugation was considered to be due only to contamination with nuclear material. But in the mid-1960s, the evidence for mitochondrial DNA began to accumulate. M. and S. Nass actually saw fine filamentous structures in mitochondria, which they interpreted as DNA, and this was later confirmed both in microscopic examinations and by very carefully extracting DNA from isolated mitochondria by chemical means.

The amount of DNA in mitochondria is quite small. A single mitochondrion may have one or more molecules of DNA, depending on its size. They are circular in shape, and in most species studied they appear to have a constant length of about five microns. This is just enough in fact to carry the coded information for the synthesis of about four or five average-sized proteins, and so the majority of the proteins in mitochondria must be made in the cytoplasm from information carried in the nucleus. It is not yet known which proteins are made in the mitochondria and which are made with the normal cellular machinery.

Both the DNA and the protein synthetic machinery of the mitochondria are different from those of the nucleus and cytoplasm in a number of features. Nuclear DNA is in the form of long filaments and is covered with a surface coating of a special protein called histone, while mitochondrial DNA is circular and lacks histone. Furthermore, mitochondrial DNA has a higher proportion of some purine and pyrimidine bases and is more resistant to structural damage by heat than is nuclear DNA. And although mitochondrial DNA can replicate in the same way as nuclear DNA, by the process described in chapter 3, it replicates at a different time in the life-cycle of the cell. In all these respects, the DNA of the mitochondria resembles that of bacterial cells more than that of the nucleus of eukaryotic cells.

The protein synthetic machinery of mitochondria also resembles that in bacteria more than that of eukaryotic cells. The ribosomes are smaller than those of the cytoplasm—more the size of bacterial ribosomes—and protein synthesis can be inhibited by certain drugs and other substances which have no effect on normal protein synthesis of the cell, but which do inhibit bacterial protein synthesis.

All of these similarities between mitochondrial and bacterial systems have led biologists to think that mitochondria evolved from bacteria present in primitive cells as intracellular parasites. This kind of situation is quite common today. For example, some single-celled animals contain small single-celled photosynthetic plants within their cells, and the animal can be kept alive in the absence of an external food supply by using some of the nutrient molecules made by the plant. In some cases the plants have become so specialized that they are unable to live outside the animal cells, so they exist together in a mutually beneficial relationship called symbiosis. Bacteria forming such a relationship with primitive cells could, so the theory goes, have evolved into the present-day mitochondria.

Plant plastids

All plant cells, with the possible exception of some algae, have in their cytoplasm small membrane-bound organelles called plastids. There are a number of different types of plastid, but all are involved in some way with the metabolism of carbohydrates.

Plastids are classified, primarily on the basis of colour, into two groups. The leukoplasts, as the name suggests (Greek: *leukos*, white) are colourless and typically contain starch, oils or proteins stored in discrete particles, the granules. They are usually rod-shaped or spheroid and are found in embryonic or sexual cells and in the regions of higher plants which receive no light. They can easily be distinguished from other cell organelles by their lack of internal structures and by the substances which they contain. Those which contain starch, for example, are called amyloplasts.

The chloroplasts

The other group of plastids is the chromoplasts. These contain certain organic pigments, and the most familiar is the chloroplast which contains the green pigment called chlorophyll, a molecule which is structurally similar to the haemoglobin of mammalian blood, except that it contains the metal magnesium whereas haemoglobin contains iron. There are a number of different chlorophylls, each with a slightly different structure and with different properties. Green bacteria contain chlorophylls similar to those of higher plants, while purple bacteria contain another chlorophyll called bacterio-chlorophyll. But the function of all these different chlorophylls is the same: to trap light energy from the sun for the process of photosynthesis.

Other pigments may be present in chloroplasts, although in general their effects and colours are masked by chlorophyll. In the autumn, however, the amount of chlorophyll in cells decreases, permitting some of these other pigments to become apparent and giving leaves their characteristic red, brown and orange shades of autumn. Some chromoplasts contain red or yellow carotenoid pigments almost exclusively, and these are responsible for the colours of certain fruits and flowers. The red colour of tomatoes, for example, is due to such pigments.

The wide variety of pigments found in the plant kingdom becomes significant when considering the many different habitats of plants. Chlorophyll is able to absorb light in the red and blue regions of the visible spectrum, thus giving it a green colour (white light minus red and blue light leaves green light), and for terrestrial plants red and blue light is sufficient for photosynthesis. But when light passes through water of more than a certain depth, most of the light is absorbed, and only blue and green light is left. Therefore plants living under water must have pigments in their chromoplasts which are able to absorb light in these regions of the spectrum more efficiently than the ordinary chlorophyll. Brown algae can use the green light, and the red algae which live at greater depths, are able to utilize blue light in photosyntheis. But green algae with only ordinary chlorophyll must live in shallow water if they are to survive.

Formation of pigments in chloroplasts is in fact dependent on available light. Colourless plant cells, such as those in roots, lack chlorophyll and hence exhibit no photosynthetic activity. But exposure to light results in the formation of green pigments in the leukoplasts. Conversely, if chloroplasts are kept in the dark the chlorophyll disappears, but will reappear when re-exposed to light.

Not all plant cells have chlorophyll localized in plastids. In the blue-green algae the pigment is diffused throughout the cytoplasm or contained in small structures called chromatophores, in which the chlorophyll forms a minute central core. Photosynthetic bacteria also contain chromatophores, containing bacteriochlorophyll and varying amounts of carotenoid pigments: yellow ones in green bacteria, and red and yellow ones in purple bacteria. In those cells which do contain definite organelles, chloroplasts are of two types. The simpler ones, found in the photosynthetic unicellular organisms, have their pigments confined to dense parallel bands within the plastid membrane. And in the complex chloroplasts of the higher plants, the chlorophyll is confined to an orderly arrangement of membranous structures called grana, which are embedded in a matrix called the stroma.

The internal structure of chloroplasts was first described in 1884 by Meyer as being finely granulated—that is, they were thought to consist of a colourless matrix (the stroma) with the minute granules (the grana) embedded in it. But it was not until 1935 that Heitz succeeded in photographing chlorophyll grana in the living cells of the leaves of flowering plants. He was able to show that the grana were platelets, seen in side view as sheets, and the presence of these chlorophyll grana was made even more evident by photographing the cells in red light, which is absorbed by the chlorophyll.

But once again it was the electron microscope, first applied to the examination of chloroplasts in 1940, which showed the fine detail of these organelles. Like all plastids, the chloroplasts are bounded by double membranes, similar again to the plasma membrane, and they possess an internal complex arrangement of membranous strands, or lamella. The grana are composed of layers of lamella separated by layers of protein and lipid, but tightly compressed, so that they appear to be stacked like a pile of coins. The size of the grana varies from 0.3 to 1.7 μ, depending on the species of plant. Using modern instruments with magnifications of some 300,000 times, the surface of the grana can be seen to be covered with small bodies called quantosomes, each of which contains about 250 molecules of chlorophyll, and it is through these particles that light is absorbed and the process of photosynthesis begins.

The number and shape of the lamella vary considerably in different cells. To take a few examples, in *Euglena gracilis*, a small, single-celled fresh-water organism, there are 20 parallel lamella in each granum, the lamella being 180–320 Å thick, with 300–500 Å spaces between them. In another fresh-water organism, *Amphidinium elegans*, there are large bundles of lamella which traverse the whole section of the chloroplast. In algae where grana are absent, the lamella are long and parallel and about 30 Å thick, and in cells of spinach there are between about 40 and 60 grana, each some 0.6 μ in diameter, in each chloroplast.

The average size of chloroplasts varies from four to six microns. Their shape varies considerably, especially in algal cells. In the chlamydomonas, a fresh water algae, the large cup-shaped chloroplast almost fills the cell, while in the euglena, the chloroplasts are rod-shaped, rather like the spokes of a wheel radiating out from the centre of the cell. In the spirogyra, a filamentous algae, the chloroplast is in the form of a ribbon, twisted into a helix. The leaves of higher plants generally contain spherical, ovoid or discoid chloroplasts, but some can be club-like with a thin middle zone and bulging ends, or vesicular with a clear, colourless centre. Other shapes include branching net-like structures or just single large bodies.

18. *Different forms of chloroplasts from plant cells.* The figure at left shows the internal arrangement of the lamellae in a typical chloroplast from a flowering plant, the lamellae being condensed and packed together in some regions to form the grana. In the upper row, the variation in external appearance in some algal cells is shown. From left to right they are Chlamydomonas, Spirogyra, Chlorella and Oedogonium

In general, the chloroplasts of cells growing in the shade are larger than those growing in sunlight.

While some cells, algae for instance, usually have only one chloroplast per cell, the numbers in higher plants can vary from about thirty to several hundreds. Sometimes they are distributed evenly within the cell, but frequently they are seen packed round the nucleus or near the cell wall. In many higher plants they are not fixed positionally, but can move around within the cytoplasm. Their distribution and orientation may vary with the amount of light energy available, and they take up positions according to the variations of the light falling on the tissue. In this way their position within the cell is dictated by their function, a situation similar to that of the mitochondria.

Chemical composition of chloroplasts has been determined by isolation (centrifugation) and subsequent chemical analysis. Chloroplasts in spinach, for example, have been found to comprise about 56% protein, 32% lipid, and 8% chlorophyll, the remainder being made up of other pigments and nucleic acids. 80% of the protein fraction is of the insoluble form, bound to lipids in the form of lipoproteins, while the rest contains the enzymes involved in the biochemical reactions of photosynthesis.

Chloroplasts, like mitochondria, can be formed by division of pre-existing bodies. In fact, when a cell contains insufficient chloroplasts, more are formed by division, and when there are too many, some degenerate. In growing cells they multiply by elongating and then splitting down the middle; the total time for this division process to take place has been calculated to be about eight days. But chloroplasts can also be formed from sub-microscopic bodies called proplastids, small particles which grow to about one micron diameter and then begin to form lamella from their outer membrane. When examining young cells there can be some confusion between immature chloroplasts and immature mitochondria.

Photosynthesis

Although the mechanism of photosynthesis has been worked out in detail only recently, the existence of the process has been known for a very long time. Almost 200 years ago Joseph Priestley observed that green plants, instead of respiring in the same way as animals, 'reverse the effort of breathing'—that is, they 'breathe' out oxygen. And in 1779 Jan Ingenhousz, a Duth physicist, suggested that green plants take in carbon dioxide from the air, split it up, throw out the oxygen and keep the carbon as food material. In the years that followed there was a great deal of confusion about what actually happened in the plant cells, but in 1949 Van Neil put forward another theory: that the plant uses light to split—not carbon dioxide—but water, keeping the hydrogen and throwing out the oxygen. This theory proved to be correct and was the starting point for working out the sequence of events in photosynthesis.

As with the reactions of the mitochondria, the mechanism of photosynthesis is very complex and detailed description is outside the scope of this book. The chlorophyll of the plant cells traps light energy from the sun and transforms it into chemical energy, which is then stored in chemical bonds produced during the synthesis of carbohydrates. Carbohydrates are made by combining several molecules of carbon dioxide with several molecules of water, forming sugars and releasing excess oxygen to the atmosphere. It has been calculated that every molecule of carbon dioxide in the atmosphere is incorporated into a plant cell once every 200 years, and that all the oxygen in the atmosphere is renewed by plants every 2000 years.

Photosynthesis can be thought of as a reverse of the process of respiration as it occurs in the mitochondria. There are essentially two sets of reactions, or metabolic pathways, one of which takes place in the presence of light and requires chlorophyll, and the other which uses the products of the first, but can function in complete darkness. In the light reaction, light is absorbed by the chlorophyll and used to excite hydrogen atoms (or electrons) to a high energy state. These hydrogen atoms are then passed along a reaction chain, similar to the electron transport chain, and their energy is used to make molecules of ATP. But while the mitochondrial reaction used oxygen in a process of oxidative phosphorylation, the light reaction of photosynthesis uses no oxygen and is called simply photo-phosphorylation, and with this pathway the plant cell can make up to thirty times as much ATP as it can with its mitochondrial systems. It is during the light reaction that the oxygen is split off from the water and released.

The energy which is temporarily stored in the ATP is now used in the dark reaction to reduce atmospheric carbon dioxide, combine it with hydrogen and produce simple carbohydrates. This takes place in a cycle of reactions, somewhat similar to the citric acid cycle; the carbon dioxide combines with a five-carbon compound to give a six-carbon carbohydrate. By a series of reactions, this is split into a pair of three-carbon fragments, some of which leave the cycle and combine to form glucose, while others rearrange to reform

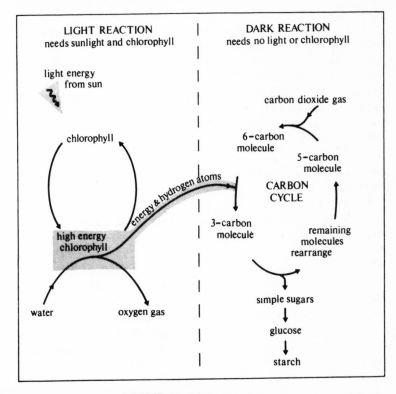

19. Green plants can store energy in food materials by the processes of photosynthesis. Light energy is trapped by the pigment chlorophyll, and used to energize hydrogen atoms split from water, oxygen being liberated at the same time. In the second part of the process, neither light nor the pigment is required. The hydrogen atoms combine with carbon dioxide so that their energy is stored as sugars, which can then form the food materials of all living cells

the five-carbon compound which combines with more carbon dioxide to start the cycle over again. This process is often called the Calvin cycle.

The first products of photosynthesis are soluble sugars. These can be joined together and stored as the insoluble polysaccharide starch (which can be seen in chloroplasts by staining them with iodine solution, when it shows up as blue particles) or other polysaccharides. These are stored initially in the chloroplast, or more commonly in the leukoplasts, and can be used as reserve food material for the cell or as structural components, for example, for making the cellulose cell wall.

Just as in the mitochondria the two sets of reactions take place at different sites in the organelle, so do the light and dark reactions in the chloroplast. The photophosphorylation stage is catalyzed by enzymes which are intimately associated with the membranes which make up the grana, while the enzymes for the dark reaction and the formation of carbohydrates are free in the stroma.

Nucleic acids in the chloroplasts

Another similarity between the mitochondria and the chloroplasts is that they both contain specific DNAs and machinery for making their own proteins. Chloroplasts have DNA strands as long as 150 μ, and special ribosomes smaller than those in the rest of the cytoplasm. It has been suggested that the chloroplasts also evolved from intracellular parasites, in this case perhaps blue-green algae, which were trapped and put to work by plant cells at the dawn of evolutionary time. But some scientists have gone so far as to suggest that both the mitochondria and the chloroplasts evolved from the same primitive structure, which in the early stages of evolution performed the energy converting functions of both organelles. Perhaps as we learn more of the cell's working, this is another question which will be answered.

6

How Cells Reproduce

When a living cell reaches a certain critical size it divides, producing two new daughter cells with the same genetic information in their chromosomes as the original cell. Cell division is the way in which single-celled organisms produce more of their kind and keep the species alive, and multicellular organisms increase their size from a single fertilized egg cell into a large and conplex adult containing many millions of cells. Cells reproduce themselves by an exact copying and splitting process, and the intricate timing and co-ordination involved make this one of the most beautiful events in nature.

Rates of cell division

The rate at which cells divide seems to be related to both hereditary and environmental factors. Each kind of cell, and cells from a particular species or tissue, has a characteristic time period between divisions, and this characteristic is passed on from generation to generation in the genetic information. The major environmental factors are prevailing temperature and the availability of food materials. For example, at 38°C cells in the nerve tissues of embryonic grasshoppers have been found to divide one every three-and-a-half hours, but if the temperature is lowered to about 18°C, this period is lengthened to some eight hours. Generally cell division proceeds at a slower pace in cooler conditions, and if the temperature is too low, it may cease altogether. Because cell division involves the synthesis of DNA, which requires the presence of ATP and enzymes, it is dependent on the environment for supplies of raw food materials to make them, and if such materials are in short supply the cell's metabolism slows down and the rate of cell division decreases.

Cell division in multicellular organisms is at its most active in new and rapidly growing tissues, and in situations where cells are being replaced in mature structures. But among the most rapidly dividing cells are the bacteria; some of the rod-shaped bacteria living in the large intestines of mammals can double in size and weight and divide every twenty minutes when they are growing at their fastest rate. Starting from a single bacteria, after twenty minutes this will divide to produce two cells, each of which will divide twenty minutes later to increase the population to four cells. Thus every twenty minutes the population doubles, so that after ten divisions—which would only take 200 minutes—there would be 1024 cells, and after just twenty-four hours and seventy-two divisions, one single bacterial cell will theoretically have produced a mass of bacteria weighing some 5000 tons. But this figure can only be theoretical, as the supply of food material would have been exhausted long before the population reached these proportions, and cell division would have slowed down and stopped.

Sexual and asexual reproduction

Reproduction in living organisms is generally of two types, sexual and asexual. In asexual reproduction, which is the form seen in simple organisms such as the amoeba, the cell simply splits into two parts, each daughter cell having identical chromosomes with the same genetic message as the parent, and a share of the cytoplasm and the cytoplasmic organelles sufficient to enable it to exist as a separate entity and to grow and divide itself at some later time. It is also by this method, called mitosis, that mature cells divide to increase the size of a multicellular organism. Sexual reproduction is vastly more complicated. Two special different cells, one from a female and one from a male of the species must fuse and the chromosomes and the cytoplasm mix together to give a mixture of the parental characteristics. These special cells are called the gametes, and their production is the result of a special method of cell division called meiosis, in which the number of chromosomes in each cell is reduced by half, ensuring that the cell formed as a result of the fusion of two gametes has the correct number of chromosomes with which to start life, and that the number of chromosomes in a cell is not double in each succeeding generation. This process also ensures that parental characteristics are mixed in the offspring.

Mitosis

Although the events which take place in mitosis follow closely after each other in a continuous sequence, it is convenient to consider the process in a number of phases. Before it starts, the cell is in interphase, sometimes called the resting phase, although this description is far from correct, as at this time the cell is busy manufacturing large amounts of nucleic acids and proteins in preparation for later enlargement and duplication of the chromosomes and cytoplasmic constituents. At this time the chromosomes are not visible within the cell, as they are diffused throughout the nucleus in a highly hydrated form.

The first stage of mitosis proper is called the prophase. This is marked by a gradual shortening, thickening and spiralization of the chromatin threads of the chromosomes as the DNA becomes less hydrated and shrinks. The chromosomes begin to become visible bodies which can be stained with Feulgen stain, and as prophase proceeds they become even shorter and thicker and can be stained more deeply. By the end of prophase they have shrunk to short, rod-like structures, only about one twenty-fifth of their length at the start of prophase, and they have all moved to the middle region of the nucleus. By this time also, each chromosome appears as a double stranded structure, the second strand being the new chromatin material formed by the replication of the DNA of the original chromosome during interphase. Each of these strands is called a chromatid, and the two chromatids of each chromosome are joined together at one point by a special constriction called the centromere.

As these changes are taking place in the chromosomes, other events are happening in the rest of the nucleus and outside in the cytoplasm. As the cell is no longer actively synthesizing proteins, it does not need a supply of ribosomes and RNA and so the nucleolus where these molecules are made diminishes in size and becomes less distinct, and by the end of prophase has disappeared altogether.

Also towards the end of prophase, the nuclear envelope disintegrates, not to reappear until the mitotic process is complete.

But before the nuclear envelope breaks down, however, the centrosome adjacent to its outer membrane in the cytoplasm divides and becomes active, and star-like structures called astral rays, or simply the aster, can be seen radiating out from it. The two pairs of centrioles then begin to move round the nucleus in opposite directions so that each eventually lies opposite the other on the far side. As they do so the astral rays between them are stretched across the nucleus to form a structure called the spindle, sometimes called the achromatic structure because it cannot be stained. The general form of the spindle is of two cones placed base to base, and the fibres of the astral rays converge at the centrioles to form mitotic centres or poles. So when the cell eventually divides, each daughter cell will contain a pair of centrioles, which will be ready for the next division process. The fibres of the spindle are formed of complex proteins made in the cytoplasm, and examination with the electron microscope has revealed that they are themselves made up from much smaller fibres called microtubules. These are like hollow tubes, about 200 Å to 270 Å in diameter with walls some 50 Å to 70 Å thick, and they lie in straight parallel bundles to form the spindle fibres. In plant cells which have no centrosome, the spindle forms a little later in prophase, after the nuclear envelope has broken down and disappeared, and the mitotic poles are merely specialized regions of the cytoplasm.

When the spindle is complete, the chromosomes have moved to the region midway between the poles—the area called the equator of the spindle or the equatorial plate—and the centromere of each pair of chromatids is attached to one of the fibres. When all the chromosomes have lined up and become attached, the cell is said to be in the next stage of the mitotic process, the metaphase. This is the shortest phase of all; no changes occur in the cell, and the name is used merely just to describe the state of the chromosomes on the spindle. But the shape and appearance of the chromosomes at this time is used to classify them into four types, according to their shape. Chromosomes which are rod-like with the centromere at their proximal end are called telocentric. If they are rod-like, but with a very short, almost imperceptible arm attached to the other side of the centromere, they are called acrocentric. Chromosomes with two definite arms in the shape of an L, the centromere being at the join of the two arms, are called submetacentric. And if there are two equal arms on either side of the centromere, with the whole chromosome in the shape of a V, it is called metacentric. This classification of chromosome shape can be very important to biologists, as the number and shapes of the chromosomes in a species can be a guide to identification.

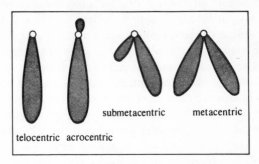

20. During mitosis chromosomes shorten and thicken until they are seen as short rod-like structures. During metaphase, four characteristic shapes are recognized, depending on the position of the centromere

The next stage, called anaphase, begins as the centromere of each chromosome divides, and the chromatids begin to separate and move towards the mitotic poles. This division of the centromeres takes place apparently simultaneously in all the chromosomes. Each chromatid is drawn towards its mitotic pole by a fibre of the spindle, and the way in which this happens still puzzles biologists. Some think that the chromatids are pulled from the equatorial region by contraction of the spindle fibres, which seems to be a very reasonable explanation, as the fibres are made of protein which, as we have seen before, have contractile properties in other structures and parts of the cell. But in some cells, the chromatids appear to travel beyond the mitotic poles, which casts some doubt on this idea. Another theory is that the chromatids are pushed towards the poles as the fibres between them in the middle of the spindle elongate, but this theory is open to as much doubt as the other and the mechanism of the process is still far from clear.

By the end of anaphase the chromatids, which are now really daughter chromosomes in their own right, have clumped together at the mitotic poles at the opposite sides of the cell, and the division process has now passed into the fourth and final stage, called telophase. In some ways the events during this stage can be thought of as the reverse of those of the prophase. At each pole the chromosomes gradually lose their deep-staining characteristics and uncoil to become slender intertwined structures like the original chromatin network. Also at each pole a new nucleolus begins to form and a nuclear envelope appears, making a new nucleus segregated from the rest of the cytoplasm. Simultaneously, in animal cells, a cleavage line begins to appear round the cell in the region of the spindle's equatorial plate; as the differentiation of the nucleus proceeds, the cytoplasm in this region undergoes transformation from a gel to a sol consistency, and the cleavage line becomes a definite cleavage furrow. This gradually intrudes deeper into the cell until eventually the two halves of the cell are separated into two new daughter cells, which then revert to the interphase state and begin to synthesize the proteins and other complex organic materials necessary for life and growth. In the plant cells, the formation of the cleavage line is replaced by the appearance of a cell plate in the equatorial plate region. This is the structure which will form the cellulose cell walls of the daughter cells, and it gradually thickens until again two new separate cells are produced to begin life on their own.

Cell division, then, also involves division of the cytoplasm, and sharing of the cytoplasmic organelles. Normally the organelles are divided equally between the daughter cells, but as most of them are present in large numbers, random distribution between the new cells may lead to slight variations. Each cell, however, always receives sufficient to be able to start an independent existence, and any slight discrepancies can be made up rapidly by the new cell.

The time taken for the whole process of mitosis varies considerably from cell to cell, and can be anything from twenty minutes in some to more than six hours in others. If cells from actively dividing tissues, such as the tips of the roots and shoots of plants, are stained and examined with the optical microscope, all stages of division can be clearly seen. But it is perhaps worth emphasizing once again that the stages of mitosis are labelled only for convenience, and that they merge into each other in a continuous non-stop process.

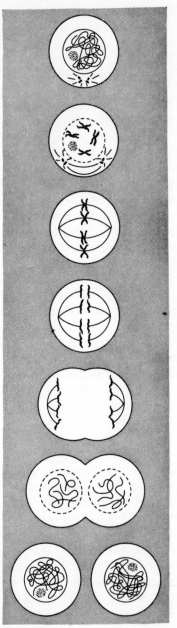

21. *Sequence of events in mitosis.* For clarity, only four chromosomes have been shown in the cell, but of course in cells with more than four chromosomes, each one behaves in the same way. From top to bottom the stages shown are interphase, mid-prophase, metaphase, anaphase and late anaphase merging into early telophase

Cells which do not divide

Although some cells, such as those in the ovaries and the testes of animals, seem to exist only to divide and produce more cells, some other cells, after they have reached full development and maturity, appear to have lost the power to divide. Sometimes this power can be restored by an external stimulus. When organs are injured, for example, many types of cells may divide to produce new structures; in this way the crayfish can grow a new limb to replace a severed one and the seaweed called *fucus* can grow new 'leaves' from damaged pieces. The healing of a small wound in the skin of a mammal is another example of this same process, sometimes called regeneration. Gall production in plants is also initiated in response to an external influence, the stimulus here being a substance injected into the plant by an insect. The plant reacts to this stimulus by commencing rapid cell division, resulting in the small round growths so familiar on oak trees and other plants.

Some cells are incapable of dividing at all, as they contain no living substance. Robert Hooke's cork, and indeed the corky tissue in the bark of all trees, is a good example of such a tissue which has lost its living matter. Other cells, although quite alive, have never been known to divide once they have formed and matured, and these include the nerve cells of animals, which spend their whole life in the interphase condition.

The specialized reproduction cells

In sexual reproduction, male and female reproductive cells—the gametes—fuse together to form a zygote, the single fertilized cell from which a mature organism will eventually develop. But if the gametes were formed by mitosis, so that they contained exactly the same chromosomes as their parent cells, the zygote would contain twice the normal number of chromosomes, and at each succeeding generation, the number would double again. It was suggested in 1887 that in each generation of living organisms there must be a stage at which the number of chromosomes is reduced by half to form the gametes, and later in the same year Flemming observed that just before the formation of eggs and sperms in animals, there was a division which differed from mitosis in that two nuclear divisions followed in rapid succession. This was the first direct observation of the process called meiosis, or sometimes simply reduction-division. It is achieved by the chromosomes dividing once while the rest of the cell divides twice, so that one cell with the normal number of chromosomes gives rise to four cells, each with only half the normal number of chromosomes.

Every cell of a particular species, plant or animal, has a specific number of chromosomes. This is almost always an even number, as the cell has inherited one half of the complete set from its maternal parent and the other half from its paternal parent. This normal number of chromosomes is called the diploid number; in man it is 46, in a horse 66, in a sunflower 34 and in a garden pea 14. Furthermore, each chromosome in a cell has its 'double', or homologue, another chromosome which, although genetically different, appears very similar in structure—again, because it has inherited one of each pair from each of its parents. After meiosis, each of the four daughter cells contains only one homologue of each pair of chromosomes, and so each gamete is said to contain the haploid number of chromosomes. Usually the haploid number is denoted as n and the diploid number as $2n$.

Meiosis

The two divisions of meiosis are called the first and second meiotic divisions; although each is a continuous process, as in mitosis, each is conveniently broken down into stages with the same names as those in mitosis. The prophase of the first meiotic division is much more complicated than the prophase of mitosis, and is itself broken down into convenient stages. During leptotene—the first stage of prophase—the chromosomes become visible as long thin threads which can be stained with Feulgen stain. But they are seen as simple threads, and are not longitudinally split into chromatids at this stage. Also the chromomeres—the regions on the chromosomes which appear as small swellings, stain deeper than the rest of the threads, and are equivalent to the bands seen in the giant chromosomes of the fruit fly—are very clearly visible at this time. (They are not so evident during mitosis.) At the next stage, zygotene, the maternal and paternal chromosomes come together in their homologous pairs and lie very close together, so that the centromeres and the corresponding chromomeres are exactly aligned next to each other. There is now a haploid number of chromosome pairs, called bivalents. The chromosomes then begin to shorten and thicken and become spiralized, and can be stained even more deeply. As the chromosomes shorten even more, the process moves on to the next stage called pachytene (which means thick ribbon), and now each chromosome of the homologous pair can be seen as two strands, joined at the centromeres, so that the whole bivalent appears as four strands twisted together. The chromosomes are now in the same stage as when they first appeared at the beginning of mitosis, except that they are in pairs.

The fourth stage of the first meiotic prophase is called diplotene, and now the forces holding the split chromatids together seem to weaken, and the chromatids have a tendency to move apart from each other. But they still appear joined together at a number of points along their length, besides at the centromere, and up to twelve such points have been observed in some cells. At these points the chromatids break, and the broken part of each chromatid joins on to the end of the other chromatid, so that an exchange of chromatid material between the pairs of homologous chromosomes has taken place. To give a simple picture of what happens, if two pieces of tape, one red and one blue to represent chromatids from each of the chromosomes of a homologous pair are stuck together in the form of a cross, and then the cross cut down the middle, a pair of V-shaped pieces of tape is produced, each with a blue arm and a red arm, so that an exchange of tape has occurred between the two original pieces. This mechanism is called crossing over, and is an extremely important part of meiosis, as it ensures that each gamete eventually produced will have genetic material which is slightly different from that of the parent cell; hence the organism resulting from fertilization of these gametes will not be exactly the same as its parents. This process also ensures that any species will be made up of a variety of different individuals, although these will share common characteristics.

When the crossing over is complete, the chromosomes become even shorter and thicker and move further away from each other, and this marks the last stage of the first meiotic prophase, the diakinesis. By now the nuclear membrane and the nucleolus have disappeared, and the first meiotic spindle is being formed, the centrioles being

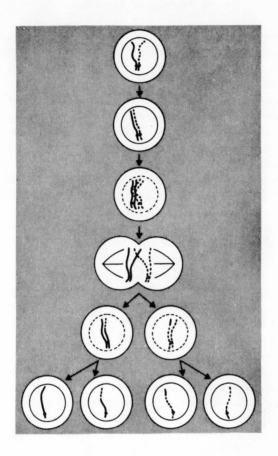

22. *Sequence of events in meiosis.* In these diagrams, for clarity, only one pair of chromosomes is shown. The stages shown are, from top to bottom, late leptotene, zygotene, diplotene (crossing over), diakinesis, the end of the first meiotic division and the end of the second meiotic division, where the number of chromosomes has been reduced by half in each daughter cell. The intervening anaphase and metaphase stages have been omitted, as they are similar to the same stages in mitosis

involved here in animal cells just as in mitosis. When the spindle is complete and the meiotic poles have been set up, the chromosomes line up on the equator with the centromere of each chromosome (which still consists of two chromatids) attached to the spindle fibres. This is the metaphase, and it is swiftly followed by the anaphase, when the centromeres move to the poles of the spindle, pulling their chromosomes behind them. Unlike anaphase in mitosis, the centromere does not divide at this stage, but remains intact joining the two chromatids of each chromosome together. Eventually the chromosomes reach the poles, and the final stage of the first meiotic division, telophase, begins. In some cells there is a complete telophase with the formation of a complete nucleolus and nuclear envelope and a subsequent short period of interphase, but in other cases the nuclei of both daughter cells begin the second meiotic division almost immediately. However, in both cases, the cells now contain the haploid number of chromosomes, each chromosome being split longitudinally into two chromatids.

If the daughter cells proceed straight to the second meiotic division, the second meiotic prophase is eliminated, the only exception being that a new spindle must be formed as the first spindle has by now completely disappeared. The second meiotic spindle lies with its equatorial plane at right angles to that of the first meiotic spindle, and as soon as it is formed, the chromosomes line themselves up on it, with their centromeres attached to the spindle fibres. The rest of the process—the second meiotic metaphase, anaphase and telophase—appear at first sight to be the same as in mitosis, although there are two important differences. The two chromatids of each chromosome are not identical as they are in mitosis, because they have had parts of their threads changed as a result of the crossing-over process. And the cells now taking part in division have only the haploid number of chromosomes. When the final telophase is complete, there are four cells, each with only the haploid number of chromatids—now the new chromosomes—and each with interchanged genetic material.

The haploid cells resulting from meiosis form the reproductive cells of the organism—the gametes—after undergoing slight modifications. In males, each daughter cell from meiosis grows a flagellum or tail and becomes a motile sperm, containing a haploid set of chromosomes, but little else. In females, three of the four meiotic daughter cells are very small bodies called polar bodies and are discarded, while the fourth is much larger, contains a great deal of cytoplasm as well as the haploid set of chromosomes, and will form the egg cell. When a male and a female gamete fuse at fertilization, the resulting zygote contains a mass of cytoplasm with which to start an independent life, and a diploid set of chromosomes. And it is then ready to start dividing by mitosis and to go on dividing to produce a new adult organism.

In every living organism, meiosis takes place before the formation of the gametes, so that the gametes are always, without exception, haploid and the zygote after fertilization is diploid. But the time interval between meiosis and the production of mature gametes shows considerable variation. In nearly all animals, gamete formation in the cells of the reproductive organs takes place almost immediately after meiosis. But in many plants there is a definite, and sometimes quite long, period between meiosis and the maturation of the daughter cells into gametes. In these cases, the haploid cells produced from meiosis undergo mitotic division and produce a specialized generation of the plant in which all the cells are haploid. Gametes now arise from the cells of this special generation by simple mitosis, and they will be haploid because the daughter cells produced by mitosis have exactly the same chromosomes as the parent cell. In some of the lower plants, such as the mosses, the haploid generation is a definite plant structure consisting of many cells, although it is quite distinct from the normal diploid generation plant. But in the higher flowering plants, the haploid generation is much reduced, and is much more difficult to identify.

Sex chromosomes

It was said earlier that each cell of an organism has a characteristic number of chromosomes, and that every chromosome has its structural double or homologue. But there is a very important exception to this rule. In man, for example, each cell has 46 chromosomes, or 23 pairs. In the male there are 22 matching pairs, called autosomes and two unmatched chromosomes called heterosomes, and this unmatched pair of chromosomes is called the sex chromosomes. The larger of the two heterosomes is known as the X chromosome while the smaller is called the Y chromosome, and it is the Y chromosome which makes a male what he is in genetic terms. Cells of the female have 23 pairs of matching chromosomes, both sex chromosomes being of the X type. The X and the Y chromosomes can easily be distinguished with the optical microscope, and so it is a simple matter to determine the sex of an organism by examining the chromosomes of a cell in the division phase. This method has been used to foretell the sex of an unborn baby.

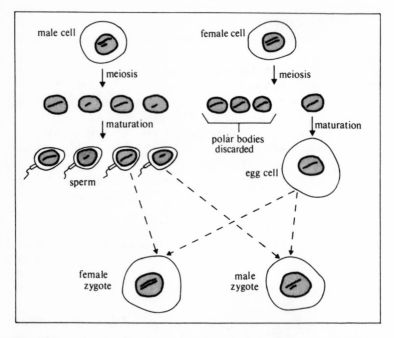

23. *Behaviour of the sex chromosomes during gamete formation and fertilization.* The male cell has two different sex chromosomes, the shorter one being the Y, or 'male' chromosome. Each male cell therefore produces four gametes, the sperms, two of which contain the Y chromosome and two the X chromosome. The female cell, on the other hand, produces just one egg containing an X chromosome, the other three daughter cells from meiosis being discarded as the polar bodies. When a male and a female gamete fuse at fertilization, producing a zygote, there is an even chance of the zygote containing either two X chromosomes or one X and one Y chromosome, an equal chance of a female or a male zygote respectively

After meiosis, all the gametes produced by the human female contain an X chromosome, but in the male, half of the sperms will contain an X chromosome and the other half a Y chromosome, so that on fertilization, if a Y-carrying sperm fertilizes an egg, a male offspring will result, and if a X-carrying sperm fertilizes the egg, the zygote will contain two X chromosomes and will develop into a female.

Sex determination of this type is the commonest form among animals. But in some cases, birds and reptiles for example, it is the female which has the heterosomes—the X and Y chromosomes—while the male has two matching X chromosomes. This means that the males can produce only one type of sperm, but the female can produce two types of eggs. Some plants also have distinct X and Y chromosomes, although many plants lack sex chromosomes and have a different mechanism for determining sex.

7

Bacteria and Viruses

The structures and characteristics of prokaryotic bacterial cells have so far been mentioned only where they differ markedly from those of eukaryotic plant and animal cells. So in the first part of this final chapter, the bacteria will be considered as cells and organisms in their own right, and their special forms and structures will be described in more detail. Of the three classes of bacteria, only the eubacteria—or so called 'true' bacteria—will be considered, as these represent typical bacterial organization. The latter part of the chapter will be devoted to a brief look at organisms which are very unlike all the other members of the living world which have been described so far: the viruses.

The diversity of bacteria

Bacteriologists have a number of ways of classifying and identifying bacteria as they see them with the optical microscope. The first is shape and arrangement of bacterial cells and three shapes are commonly recognized. The cocci (singular—coccus) are spherical or ovoid cells, usually about one micron in diameter, and—depending on the species or group of species—the cells tend to aggregate in characteristic patterns. The micrococci are usually seen as single individual cells, diplococci normally group together in pairs, the streptococci form themselves into long rows which look like chains or strings of beads, and the staphylococci are seen in regular groups rather like bunches of grapes. All of these arrangements are determined by how the cells divide and how, or if, they separate after they have divided. When trying to identify a specimen of bacteria, the shape and grouping of the cells is one of the first signs the bacteriologist looks for.

The second group are the bacilli (singular—bacillus), and these are cylindrical cells, normally referred to as rods. They are usually about one micron in diameter and vary from between five to eight microns long. Although in some species the cells are in pairs or short filaments of cells arranged end to end, the arrangements are not so characteristic and hence not so important in identification as with the cocci.

The members of the third group of the eubacteria are the helically-coiled bacteria, called spirilla. These are several times as long as the bacilli—some of the largest are as long as 20 μ—and are almost always seen as individual cells.

Experimental methods for studying bacteria

Observation and examination of bacterial cells is greatly assisted by the use of special staining techniques, many of which are similar to those used to examine plant and animal cells microscopically. But there is one staining method which is of particular importance in the identification of bacteria: the Gram stain, developed in Denmark by Christian Gram in 1884. In its original form, this method was used to make bacteria stand out against surrounding tissue, and the method was to fix the sample of tissue by heating over a flame, and then stain with crystal violet, followed by iodine. This mixture stained both the tissue and the bacteria, so Gram added pure alcohol to the preparation, which decolourized the background tissue, but left the bacteria stained bright violet. Finally he added another stain called Bismark Brown, and the final result was a brown tissue background and violet bacteria. But it was noticed that some bacteria were also decolourized when the alcohol was added, and so appeared brown in the final preparation. Gram's method has therefore since been modified to provide a method for distinguishing between bacteria which lose their stain when organic solvents are added to stained preparations and those which retain their stain. The modern method is initially to use a purple stain followed by iodine, decolourize with acetone, and then finally to add a red dye to stain the decolourized cells. Bacteria which retain the original purple stain are called Gram positive bacteria, and those which lose the dye—and hence take up the red counterstain—are called Gram negative. This differential staining reaction, apart from being an extremely valuable aid to identification, also tells the bacteriologist a great deal about the cell's structure.

Bacterial cell walls and capsules

Eubacterial cells are surrounded by a rigid cell wall, outside their plasma membrane, and this confers a definite shape on the cell. It can be removed more or less intact by rupturing the cell and permitting the contents to escape, and it will still retain its characteristic shape. It appears to have only protective and mechanical functions and to play no role in the transport of materials into or out of the cell or in metabolism. If it is digested away with enzymes or chemicals, then lysis of the cell ensues and the cell dies. The thickness of the cell wall varies considerably. In Gram positive bacteria it is about 150 Å to 200 Å thick, and seems to be connected physically to the plasma membrane over its entire inner surface. On the other hand, the cell walls of Gram negative bacteria are much thinner, only 75 Å to 120 Å thick, and are attached to the plasma membrane at only a few sites.

The first chemical analysis of the bacterial cell wall was successfully carried out by Salton in the 1950s, and its detailed structure only elucidated in the 1960s. It has been found to be much more complex than the cellulose cell wall of the plant cell. All bacterial cell walls contain large complex molecules made up of two types of sugar molecules and three or four amino acids, and the complex is called generally mucocomplex. The two sugar molecules are joined together alternately in long polysaccharide chains and these chains are arranged side by side and fairly close together. The amino acids, which are usually different from those found in the proteins in other cellular structures, are formed into short chains, and these attach to the polysaccharide chains and form cross links, so that the whole structure looks rather like basket-work with the polysaccharide chains laid one way and the amino acid chains crossing them and joining them together.

Cell walls in Gram positive bacterial cells contain large amounts of this mucocomplex which often forms the whole structure, although there are sometimes small amounts of other simpler polysaccharides present. Gram negative cell walls, however, contain much less mucocomplex, but large amounts of proteins, lipids and other polysaccharides. The lipid content may be as high as twenty to thirty per cent, and photographs of these cell walls taken with the electron microscope indicate that they may be made up in layers.

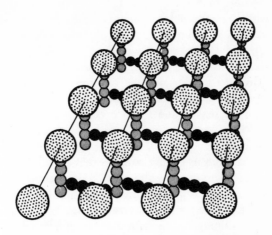

24. *Molecular structure of the bacterial cell wall.* The large spheres on the top of the structure represent the chains of polysaccharide material, which lie very close to each other. They are joined together to make a very stable criss-cross structure by short chains made up of amino acids. Together these polysaccharide and amino acid chains form the complex called mucocomplex, the amount of which in Gram positive and Gram negative bacterial cell walls varies and provides the reason for this different staining reaction

These differences in structure and composition of the Gram positive and negative cell walls provides the explanation of the different reactions to the Gram staining method. The organic solvents used are able to dissolve lipids, so that they can dissolve their way through the lipid in the cell walls of the Gram negative bacteria and penetrate the cell to leach out the dye. But walls of the Gram positive bacteria with no lipid in their structure present a barrier preventing entry of the solvent.

Outside the cell walls of many bacteria is another structure, more diffuse and less rigid than the cell wall, and forming an envelope round the cell. This is the capsule, and its size and form varies in different species and seems to be influenced by the environment in which the bacteria lives or is grown in laboratory culture. Sometimes it is a barely detectable layer, and sometimes a mass of material greater in volume than the cell itself. Sometimes it is a rigid jelly-like layer with a definite outer boundary, and sometimes there is no definite boundary and the material seems to be diffusing out from the cell wall. In fact it is sometimes difficult to know if a capsule exists at all, as it would have to be a certain thickness to be visible in the optical microscope, and the electron microscope does not help very much as the capsule may be transparent to electrons, or may shrink in the drying and fixing process necessary for electron microscopy.

The capsule is made mainly of carbohydrates in the form of polysaccharides, such as cellulose. The exact chemical composition varies from species to species, and even from strain to strain in the same species, so that different strains can be identified by determining the composition of their capsules. Its function is not completely known, although there are a number of accepted theories. It may act as protection, as a reservoir of stored food material or as a site for disposal of waste materials—a sort of dustbin for the bacterial cell.

Internal organization of bacterial cells

A definite nucleus as such is absent in bacterial cells; there is no nuclear envelope and the nuclear material is mixed in with the rest of the cytoplasm. It is possible, however, to assign a 'nuclear-region' to the site of the nuclear material in the cell. The presence of DNA in the nuclear region has been demonstrated conclusively by staining the cell with Feulgen stain, when only the nuclear region takes up the stain; normally there are no other large molecules associated with the DNA—no histones or other proteins as there are associated with the DNA of the chromosomes in plant and animal cells. But in rapidly growing bacterial cells there is a great deal of

RNA associated with the nuclear region as the DNA sends out its coded instructions to the rest of the cytoplasm.

The typical shape of the DNA in bacteria is that of a linear strand closed into a ring. Often there is only one single giant DNA strand, some 1000 μ long, but it is highly coiled and folded into a very small volume. With such a structure it is difficult to imagine how an equal distribution of the genetic material can be effected during cell division, as there is no counterpart of the mitotic spindle seen in the cells of higher organisms. But the DNA may be associated with the plasma membrane or the membrane of a small body called the mesosome, so that at division—which is just a simple splitting of the cell—synthesis of new plasma membrane material at the site of attachment of the DNA ensures separation of the two daughter DNA strands.

The cytoplasm of bacterial cells is also much simpler than that of eukaryotic cells. Because they divide so rapidly (it was mentioned in chapter 6 that bacteria can divide once every twenty minutes under optimum conditions), bacterial cells would be expected to have large numbers of ribosomes, all working at full production, and indeed in some cells nucleoprotein accounts for about thirty per cent of the dry weight, and RNA for some twenty per cent. Electron microscope pictures of dividing bacterial cells shows that the cytoplasm is extremely granular, these granules being the ribosomes, which are much smaller than the ribosomes of eukaryotic cells. They are not independent particles, but are organized into complicated groups and strings, similar to the polysomes of animal and plant cells.

Organelles in bacterial cells are few. There is no endoplasmic reticulum, Golgi complex or mitochondria, and the chief membrane-bound structure is the mesosome, which is well developed in the Gram positive cells, but less so in Gram negative cells. The membrane of the mesosome, which is of the same general structure as all other membranes, is continuous with the plasma membrane and appears to be a pocket or invagination of it. It is the site in the bacterial cell where the reactions of the electron transport chain take place and it therefore contains the enzymes associated with this process, so that it is the counterpart of the eukaryotic mitochondrion. But because it is so closely associated with the nuclear material of the bacterial cell, it does not need to contain any DNA.

Bacterial cells which are capable of photosynthesis contain another membrane-bound internal structure called the chromatophore, which although much simpler than the chloroplast of green plants, is nevertheless the site of all the photosynthetic reactions and processes. The single membrane of the chromatophore is also continuous with the plasma membrane, and the whole structure may be as large as 1000 Å across. In actively photosynthesizing bacterial cells, the cytoplasm seems to be filled with chromatophores and they even seem to obscure the presence of the ribosomes.

Chemical analysis of isolated chromatophores shows that they are made up of proteins and lipids, and that they contain a special form of chlorophyll called bacteriochlorophyll, together with other pigments such as carotenoids, all collected into a minute core within the outer membrane. They also contain the enzymes necessary for the reactions of photophosphorylation. But the actual process of bacterial photosynthesis is quite different from the process as carried out in green plants. While plants split off hydrogen from water, producing oxygen as a result, bacteria take their hydrogen from hydrogen sulphide—the well known gas which smells of bad eggs—or from some other form of organic compound. Carbon dioxide is usually used as the source of carbon, so that the green bacteria use this and hydrogen sulphide as their raw materials, and produce sugars, water and inorganic sulphur as by-products. Another difference is that photosynthetic bacteria grow in anaerobic conditions (in the absence of air) and hence they are usually found at the bottom of stagnant ponds.

Spore formation

Bacterial cells have a number of means of protection against unfavourable external conditions, such as their thick cell walls and capsules, and some species can live under conditions which would quickly kill other organisms. For example, different bacteria can live and reproduce under widely differing temperatures: some can exist at temperatures approaching freezing-point, and at the other end of the scale, there are bacteria which are quite happy in hot springs with temperatures as high as 92°C. A few species of rod-shaped bacteria have another, rather special, means of survival. Under adverse nutritional conditions they cease normal cellular activity and make an extremely resistant coat round themselves to

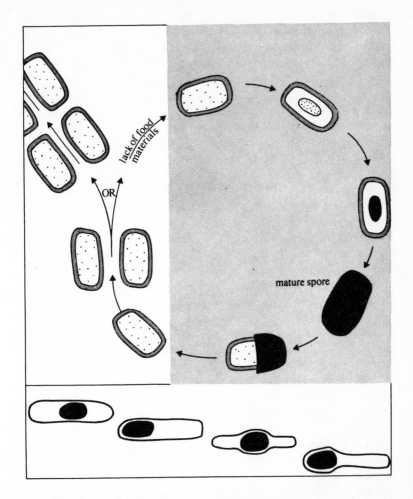

25. Under unfavourable conditions, such as lack of food materials, some bacterial cells will form spores (*top*). Nuclear material and a little cytoplasm is surrounded by a thin wall, which gradually thickens and hardens until the mature spore remains. Spores can withstand very harsh conditions for very long periods of time, but when the environment becomes more favourable they germinate to produce a new normal bacterial cell. Spores can show a wide variety of shape within a bacterial cell (*below*), although each species normally gives rise to spores of the same shape and in the same position

form a resting stage or spore. These spores, although metabolically inert, can remain viable for years—sometimes even centuries—at normal temperatures and in a dry state, and some are so resistant that they can survive being boiled for some two hours.

The process of spore formation, called sporulation, is an orderly sequence of events, beginning when the genetic material separates from the rest of the cytoplasm by the formation of a thin transverse wall within the parent cell. The wall grows thicker and surrounds the nuclear material and a little cytoplasm to form the spore core, and then various other layers are formed on its outside. First a thick cortex, made of a polysaccharide and amino acid mucocomplex similar to that of the bacterium's normal cell wall, is laid down, and then layers of protein material form on the outside of this. During these processes the spore gradually loses water, and while a bacterial cell contains some eighty per cent water, a mature spore contains only about sixty per cent. The appearance of a spore in a cell varies with different species. It may be formed in the middle or at the ends of the mother cell, and may have a diameter smaller or larger than the cell, sometimes giving the cell a bulging middle or end. Eventually the mother cell dissolves away, and the free spore is released.

Spores can revert to the normal bacterial form by a process of germination, which can be initiated by the presence of very simple nutrients such as amino acids. The cortex rapidly disintegrates, and calcium and mucocomplex leak from the spore so fast that it can lose as much as thirty per cent of its weight in thirty seconds. The spore core then forms an outgrowth through what is left of the cortex and other coats, and this outgrowth undergoes cell division, to begin a new generation of bacteria.

The discovery of viruses

By the end of the nineteenth century it was generally accepted, thanks to the work of great scientists such as Louis Pasteur, that many diseases were caused by micro-organisms such as bacteria and some single celled animals and plants. Bacteria produce disease by entering the tissues of an organism and damaging them by producing poisonous chemicals called toxins; the causative agents of many diseases, and the toxins involved, were known. But in 1892 Iwanowsky discovered another type of disease-producing micro-organisms. While studying mosaic disease of tobacco plants, he found that the juice which he extracted from the leaves of diseased plants could cause an infection in other plants. Suspecting that the infective agent contained in his juice was a bacterium, he tried to isolate it by filtering the plant juice through a very fine filter—the normal method at the time for isolating bacteria from tissues and liquids. But Iwanowsky failed to separate any bacteria, even using the finest filters then available, and he concluded that the agent of tobacco mosaic disease was smaller than any bacteria then known.

During the next few years other diseases of both plants and animals were found to be caused by organisms which passed through fine filters, and these agents became known as filterable viruses, or simply viruses. Virus means a sort of living venom or poison, and the name was used because it was already known that bacteria induce disease by producing poisons. These viruses could not be seen even with the highest available magnifications of the optical microscopes, could not be grown on the usual media used for growing bacteria in the laboratory, and retained their infective properties even after being precipitated from acid solutions which would kill bacteria and other organisms. They certainly presented a baffling problem. Then in 1935 Stanley demonstrated yet another unique characteristic of the viruses. He managed to crystallize the tobacco mosaic virus, thus proving that viruses were not cellular in the normal sense. Furthermore, the virus crystals still retained their infective capacity and could cause the symptoms of tobacco mosaic disease when injected in solution into healthy plants.

Structure of viruses

Today, modern techniques and instruments—including, of course, the electron microscope—have been applied to the study of viruses, and much more is now known of their structures and way of life. They range in size from 100 Å to 3000 Å, the longest being about the same size as a small bacterium. Although they cannot strictly be classified as cells, as they lack cytoplasm and a limiting plasma membrane, they do possess the fundamental properties of life—they have genetic material and are able to reproduce. But they can only proliferate inside living cells, and for this reason they have been described by some people as 'genetic material in search of a living cell in which to reproduce'.

Early chemical analysis of virus material showed that they were composed only of protein and nucleic acids, and it is now known that the protein forms a coat or outer envelope surrounding a core of nucleic acid. The whole virus particle is conventionally called the virion and the protein coat the capsid. The capsid itself is composed of a number of sub-units of characteristic shape which are referred to as the capsomeres.

Unlike the cells of plants, animals and bacteria, viruses contain only one type of nucleic acid—either RNA or DNA—and so far no virus has been found to possess both. Generally the viruses which invade and reproduce inside plant cells contain RNA as their genetic material, and of those which invade animal cells and bacterial cells—the latter are called bacteriophages—some contain RNA, but the majority DNA. Some viruses and bacteriophages are very specific in their choice of host cell, and will only invade and reproduce in one type of cell, while others will attack many different cells. They can be grown for study in the laboratory under carefully controlled conditions inside fertilized hens' eggs or in cells in tissue culture, by methods and under conditions rather similar to those which must be used to grow plant and animal cells.

Viruses can be seen and examined directly with the electron microscope, and their shape is seen to vary considerably. Specimens for electron microscopic examination are 'stained' with heavy metals, evaporated on to the particles from the side to give a 'shadowing' effect, so that they are seen in photographs as light objects against a dark electron-dense background. Plant and animal viruses are normally rod-shaped or polyhedral, and have much simpler structures than the bacteriophages which will be described later. The tobacco mosaic virus is a simple rod about 3000 Å long and some 180 Å wide. It was the first virus to be taken apart and reassembled in a test tube, and it was found to be composed of a long RNA molecule surrounded by a cylindrical capsid. The capsid is made up of 2130 capsomeres, all made of the same protein, and arranged in a helix with sixteen of the capsomeres in each turn of the helix. A simple model of this virus can be constructed by wrapping a string of closely threaded beads round a finger, the beads representing the helical capsid and the finger re-

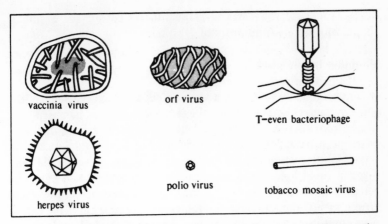

vaccinia virus

orf virus

T-even bacteriophage

herpes virus

polio virus

tobacco mosaic virus

26. *Examples of the structures of some virus particles, drawn to scale.* Because the nucleic acid core of the virus must contain all the coded information for the synthesis of the protein coat and for the way in which the coat is assembled, the more complex viruses, such as the T-even phage, must contain much more complex nucleic acid than a simple virus such as the tobacco mosaic virus

presenting the RNA core. The virus responsible for influenza is also seen to be of the same structure, although with a different number of capsids of different proteins.

Many virus particles appear at first sight to be spherical, but greater magnification shows that they are regular polyhedrons. The adenoviruses, which cause a type of respiratory disease in man, are icosahedrons, shapes with twenty regular faces, and are composed

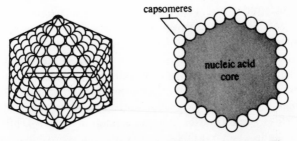

capsomeres

nucleic acid core

27. *Internal and external structure of the adenovirus.* The protein capsid is made up of 252 individual capsomeres, which form a coat around the nucleic acid core

of 252 spherical capsomeres; many other viruses have a similar shape, with different numbers of capsomeres.

Viral infection of plant and animal cells

It is not known exactly how a virus enters a living cell. Some of the smaller, simpler ones enter by phagocytosis, and this has been seen in electron microscope photographs. Some viruses have an outer envelope surrounding the whole virion, and this could fuse with a cell's plasma membrane. Enzymes present in the capsid could then digest a small hole in the membrane and the virus could enter the cell through the hole. Nor is it known in all cases whether the whole virion, or just the nucleic acid core, enters the cell, but if the capsid does enter it is soon dissolved away, leaving the naked nucleic acid. In all cases, however, the process is quite rapid, and takes only a few minutes.

Once inside the cell, the viral nucleic acid immediately overpowers the resident genetic material and takes control of the cell's stores of raw materials, its metabolic machinery and its energy producing systems. It then directs the cell's protein synthetic machinery to make proteins according to its own coded information, and these proteins will eventually be used to make the protein coats of new virions. Because a virus can contain only one type of nucleic acid, either RNA or DNA but not both, viral RNA must be as capable of carrying genetic information and of directing protein synthesis as normal DNA.

Double stranded viral DNA in a host cell's nucleus first undergoes duplication by the usual mechanism, drawing on the cell's supply of free nucleotides and using the cell's own enzymes to catalyze the process; viruses frequently contain no enzymes except those used to gain entry to the cell. One strand of the viral DNA is then used as the template for synthesis of a viral form of messenger RNA, again using raw materials from the cell's own reserves. This viral messenger RNA contains the information for the synthesis of viral coat proteins, which are then made in the cytoplasm using the cell's existing ribosomes, transfer RNAs and amino acids; while these are being made, the viral DNA is busy making more copies of itself in the nucleus. Eventually the new coat proteins will assemble as capsomeres, and organize themselves into capsids around strands of new viral DNA to make a large number of complete new virions. Several hundred new virions can be made in each host cell in this way.

Although most DNA viruses contain double-stranded DNA molecules in the usual double helix structure, a few contain only single-stranded DNA, which is coiled up in an irregular structure, rather like RNA in plant and animal cells. When this type of DNA enters a cell, its first task is to use the cell's resources to make a duplicate strand, and the two strands then form themselves into the usual double helix, which then produces further copies of double-stranded molecules. The normal convention is to refer to the original viral DNA strand as the + strand, and the first copy as the − strand. It is the + DNA strand, and its subsequent copies, which acts as the template for the synthesis of messenger RNA which then goes out into the cytoplasm and directs synthesis of viral coat proteins as before.

A somewhat similar situation occurs when an RNA-containing virus invades a cell. Sometimes the single RNA strand acts as the template for the production of DNA molecules, which then direct the synthesis of viral messenger RNA in the usual way. But in other cases the RNA itself acts as the self duplicating model and the template on which messenger RNA is manufactured. It is extremely interesting that in both of these cases, where single-stranded nucleic acid molecules invade cells and take over their synthetic systems, a mechanism fundamentally different from that occurring in plant, animal and bacterial cells is involved. For in normal living cells the synthesis of all DNA and RNA is carried out on a double-stranded template, and none is made on a single-stranded molecule.

Just how the viral nucleic acids manage to dominate the machinery of a cell and substitute themselves for the cells own chromosomes is still not understood. In some cases normal cellular synthesis continues even with the virus present, but in extreme cases, all RNA and DNA synthesis on a cell's chromosomes is stopped and the cytoplasmic messenger RNAs with the information for the cell's own proteins are destroyed.

When all the parts of the new virions have been manufactured and assembled so that new protein coats surround new copies of the viral nucleic acid, the host cell ruptures and the virions are released to invade and take over more cells, and produce even more copies of themselves. It is this destruction of the cells which gives rise to the disease symptoms associated with invasion by viruses. What induces the host cell to rupture is also not yet clear, although some scientists think that the release of enzymes from the lysosomes may be involved.

Viruses which attack bacteria

Bacteriophages, the viruses which specifically invade bacterial cells, are much more complex in structure than the plant and animal viruses. They began to attract interest in the late 1930s when biologists started to investigate the phages which invaded a bacteria called *E. coli*—the colon bacillus. One of the smallest of the viruses of this bacteria is called MS2, and it contains just enough RNA to provide the information for the synthesis of three proteins, the coat protein, an enzyme for making more RNA and a 'maturation factor' which assists in some way to reassemble the whole virions from new proteins and nucleic acids.

The most well known of the bacteriophages of *E. coli* are called T2, T4 and T6, sometimes called collectively the T-even phages, and these are among the most complex of all viruses. A protein coat, enclosing the DNA molecule, forms a regularly shaped 'head', and attached to this is a complex tail. The tail is a hollow tube surrounded by an outer contractile protein sheath, and attached to the bottom of this is a base-plate with many fine fibres attached to its underside. For such a complex structure, the amount of information carried on the DNA molecule must be much more than that carried on the nucleic acids of plant and animal viruses, and geneticists have been able to determine which parts of the DNA contain the information for the synthesis of the proteins, and which parts provide the instructions for their assembly into whole phage particles.

The fibres at the bottom of the base-plate of the T-even phages are used to attach the phage to the outer surface of the bacterial cell and anchor it there. Enzymes from the tail then digest small holes in the bacterial cell wall and plasma membrane, and it is through these holes that the DNA will pass into the cell. Photographs taken with the electron microscope show that the DNA is actually 'injected' into the bacterial cell through the hollow tail, the head acting as a sort of plunger, compressing the contractile tail and forcing the DNA strand downwards into the bacterium. Once inside the cell, the DNA takes over in the normal way, makes copies of itself and directing the cell's synthetic machinery to make new proteins for the head and the tail, base plate and fibres. Between 100 and 1000 new phage particles are produced inside a single bacterial cell, and when they are completely formed the cell wall is broken down and the phages released. The whole cycle is extremely rapid, taking only a few minutes between the injection of the DNA and the release of the new phages.

Temperate infection by viruses

This sequence of events is called virulent infection, and the bacteriophage is known as lytic phage, because it lyses, or breaks down the bacterial cell. But sometimes another process occurs. Instead of taking over control of the cell, the viral DNA becomes attached to the existing DNA strand of the host cell. It is then called a prophage, and appears to behave simply as part of the bacterial cell's genetic material, duplicating once every cell generation as the bacterium divides. This prophage may be carried on the bacterial DNA indefinitely and, far from destroying the cell, it may even perform some functions which are beneficial; for instance, it may give the bacterium resistance to attack by antibodies. Bacteriophages which act in this way are called lysogenic phages, and the infection referred to as temperate infection. But very occasionally—on average once in every 10,000 divisions of the bacterial cell—the prophage is released from the bacterial DNA strand and begins to multiply rapidly, and when this happens the infection becomes virulent, many new phage particles are produced, and the cell is lysed to release them.

A process similar to temperate infection of bacteria is known to occur in many plant and animal cells and tissues. In these cases, although the cells are infected, no symptoms of disease are produced, and the cells are said to be carriers of the virus or the disease. It has even been suggested that a viral DNA strand incorporated on to the chromosomal DNA of mammalian cells in a form of temperate infection can upset the normal workings of the cell and induce cancer. Certainly a relationship between viruses and cancer has

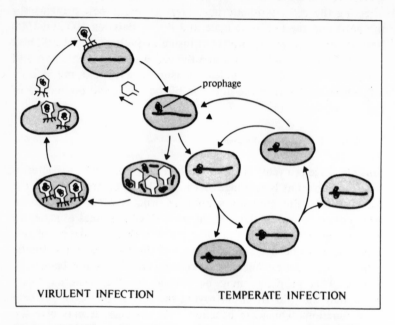

prophage

VIRULENT INFECTION   TEMPERATE INFECTION

28. Bacterial cells can be attacked by bacteriophages in two ways. In the more common virulent infection, the viral nucleic acid dominates the cell's nuclear material, and sets the cell to make new viral components which, when assembled into complete bacteriophage particles, are able to infect other cells. In temperate infection, the viral nucleic acid merely combines with the bacterial nuclear material, and behaves as though it were a part of it, even dividing with it as the cell divides. But occasionally the viral nucleic acid can become detached, and the infection becomes virulent

been suspected for some time, and as long ago as 1910 it was observed that a particular type of virus could cause cancer tumours to develop in chickens. Since then viruses have been associated with many types of tumours in mammals, particularly mice. Viruses have even been isolated from cancer tumours in humans, but because of the obvious limitations of experiments on humans, there is no conclusive proof that the viruses are causative agents.

Can viruses provide more clues to the nature of life?

Many scientists find it rather difficult to decide whether viruses are very primitive or rather specialized organisms. On the one hand, viruses contain no cytoplasm or organelles and are incapable of manufacturing their own materials or of reproducing without the help of plant, animal or bacterial cells. So they could well be called primitive. On the other hand, it is possible to regard viruses as having specialized to the stage where they do not need to carry their own metabolic factories around with them, but can live and proliferate at the expense of the more cumbersome plant, animal and bacterial cells and, in the absence of any immediate hosts, can go into a dormant but viable resting phase and wait. But whatever one's view, the study of viruses—the ways in which they are able to switch from apparently living to apparently non-living dormant or even crystalline forms, and the ways in which they can take over a perfectly normal living cell and direct its processes, systems and materials for their own ends—can perhaps provide more insight into the structure and functions of living organisms. It may even lead us one step nearer to the understanding of the nature of life itself.

APPENDIX

Suggestions for Further Reading

For those readers whose wish to study cell biology in more detail, or wish to discover more aspects of the subject, the following books will be very useful:

The Living Cell by Oliver Gillie (Thames and Hudson, 1971).
The Structure of Life by Roystan Clowes (Pelican 1967).
Cell Biology by E. J. Ambrose and Dorothy M. Easty
　　(Thomas Nelson, 1970).
A Guide to Sub-Cellular Botany by C. A. Stace (Edited by A.C.Shaw)
　　(Longman 1971).

Many articles in *Scientific American* deal with specific topics in cell biology in greater detail. Articles are available as separate off prints from booksellers, and the following are just three examples:

The Genetic Code III by F. H. C. Crick, *Scientific American*,
　　October 1966.
The Structure of Viruses by R. W. Horne, *Scientific American*,
　　January 1963.
The Living Cell by Jean Brachet, *Scientific American*. September
　　1961.

Index